Alfred James Copeland

Bridewell Royal Hospital

Past and Present

Alfred James Copeland

Bridewell Royal Hospital
Past and Present

ISBN/EAN: 9783744762311

Printed in Europe, USA, Canada, Australia, Japan

Cover: Foto ©berggeist007 / pixelio.de

More available books at **www.hansebooks.com**

EDWARD VI.
From an Oil Painting in Christ's Hospital.

Bridewell Royal Hospital

PAST AND PRESENT

A SHORT ACCOUNT OF IT AS

Palace, Hospital, Prison, and School

With a Collection of Interesting Memoranda hitherto Unpublished

By ALFRED JAMES COPELAND, F.S.A.

TREASURER OF THE ROYAL HOSPITALS OF BRIDEWELL AND BETHLEHEM

London

WELLS GARDNER, DARTON, & CO.

PATERNOSTER BUILDINGS

1888

TO

THE PRESIDENT AND GOVERNORS

OF

THE TWO ROYAL HOSPITALS

OF

BRIDEWELL AND BETHLEHEM

𝔍 𝖉𝖊𝖉𝖎𝖈𝖆𝖙𝖊

THIS BOOK.

a

PREFACE.

THE present volume is the outcome of an attempt to put into interesting shape and form, some of the old records and facts, respecting one of our ancient Royal Foundations.

After the lapse of three hundred years, notwithstanding their somewhat altered conditions, the three institutions termed the "Royal Hospitals" of Christ's, Bridewell, and St. Thomas's still remain no mean portion of the inheritance bequeathed to the citizens of London; and it is matter of congratulation, that great good is still being done by them to the community at large.

Whether for the sick and needy, or for the education of the middle classes, and the poor and necessitous, our Royal Hospitals are noble monuments in these times of the thought, care, and unselfishness of our forefathers for the benefit of others.

In the compilation of this work I have consulted all the available authorities on the subject; and to all those who have so kindly assisted me I desire to make my warm acknowledgments. I have drawn largely from Stow (Strype's edition), Pennant, and Maitland, and from "Remembrancia, or Records Preserved among the Archives of the City of London, A.D. 1579–1664."

I am indebted to Mr. Anderson of the British Museum, and Mr. Overall of the Guildhall Library, for much that is most interesting.

Such books as the "Autobiography of Thomas Ellwood," "Ned Ward's London Spy," and Mr. Dixon's work upon Prisons, have interesting and valuable allusions to the old Bridewell; and Mr. Ottley Martin's extensive Parliamentary Report, issued in 1837, has been of the utmost service to me.

I have to express my thanks to my friends the Rev. E. Rudge, the late Chaplain, and the Rev. E. C. Hawkins, Vicar of St. Bride's, for many useful hints; to Mr. St. John Hope of the Antiquarian Society, for assistance respecting the old Palace inventory of soldiers' "*Harness*," given on page 9, and Mr. Allan Barraud for the drawings of the present schools; and to the Rev. Hayman Cummings, F.R.Hist.Soc., for the valuable aid he has rendered me in the literary compilation of the work.

In these days when corporate bodies, the magnificent old Companies and Foundations of the City of London in particular, are from time to time subjected to attack from without, every effort to show their usefulness is to be welcomed; and it cannot be too widely known that these institutions are doing munificent and charitable work of the most valuable kind, in the most unostentatious way.

This book bears evidence that Bridewell, in spite of altered times and circumstances, such as never could have been even guessed at by its royal founder, is still fulfilling his intention, in the amelioration of the condition of the poor and necessitous.

A. J. C.

BRIDEWELL ROYAL HOSPITAL,
May 30, 1888.

CONTENTS.

Bridewell Hospital.

CHAPTER I.

"THE OLD PALACE."

BRIDEWELL Hospital!—What was it? What is it

ERRATA.

Page 100 (foot-note).—*For* "a mother," *read* "another". after "railings," insert "outside"; after "contempt," insert "for the place."

of our city walls, enabling them (in the words of the Rev. E. C. Hawkins, in his monograph of "The Church and Parish of S. Bride's") "to realise how this neighbourhood looked and how men lived when the great parishes of S. Margaret's, Westminster, and S. Dunstan's, Stepney, came close up to Aldgate and Ludgate, and everything outside the city walls was not London, but a suburb."

At that angle in the river Thames which was formed by the estuary of the Fleet Ditch, stood for

Bridewell Hospital.

CHAPTER I.

"THE OLD PALACE."

BRIDEWELL Hospital! What was it? What is it now?

The answer to these questions contains much that will interest those who have a taste for the old chronicles of past days and the history of the years that are gone, written in the stones and foundations of our city walls, enabling them (in the words of the Rev. E. C. Hawkins, in his monograph of "The Church and Parish of S. Bride's") "to realise how this neighbourhood looked and how men lived when the great parishes of S. Margaret's, Westminster, and S. Dunstan's, Stepney, came close up to Aldgate and Ludgate, and everything outside the city walls was not London, but a suburb."

At that angle in the river Thames which was formed by the estuary of the Fleet Ditch, stood for

A

many years the Eastern Arx Palatina or Castellated Palace of the City of London.

So far as can be ascertained, this once royal precinct occupied the space which at present is bounded by Fleet Street, New Bridge Street, and the Thames Embankment.

To the parish is given the name of the Danish Saint Bridget, and here is the only church in London dedicated to Saint Bride.

A holy spring once was here, supposed to possess miraculous curative power; it also bore the Saint's name, and in course of time the royal palace hard by was termed "Bridewell."

Difficult as it may be to realise now, there was a time when this spot was a beautiful, well-wooded retreat, separated from the busy scenes of the adjacent city by the swiftly flowing stream, bearing on its bosom the contents of the Turnmill and Oldbourne brooks, and hence called "Fleet," whilst its main bank was washed by the tidal waters of the Thames.

The discovery of relics from time to time in the immediate vicinity proves that it was known to the Romans, but the earliest records extant lead to the conclusion that the first building of repute was a tower or castle pertaining to the King. "For I read," says Stow, "that in 1087, the 20th of William I., the city of London with the Church of S. Paul being burned, Mauritius, then Bishop of London, afterward began the foundation of a new church, whereunto King William gave the choice stones of his castle standing near to the bank of the river of Thames, at the west end of the citie. After this Mauritius, Richard his successor purchased the streets about Paul's Church, compassing the same with a wall of stone and gates. King Henry I. gave as many of the stones from the walls of the castle yard as served to enclose and form the gates and precincts of the church." This tower or castle being thus "destroyed or dismantled, stood as it may seem in place whereon standeth the house called Bridewell."

" For, notwithstanding the destruction of the said castle or tower, the house remained large, so that the kings of the realm long after were lodged there and kept their courts."

" For until the 9 yeare of Henry III. the Courts of Law and Justice were kept in the King's house wheresoever he was lodged, and not elsewhere, and that this may be proved, I quote the following record :—

" Hæc est finalis concordia, facta in Curia Dom. Regis apud Sanct. Bridgid. Lond. a die Sancti Michaelis in 15. dies Anno regni Regis Johannis 7. Corā G. Fil. Petri. Eustacio de Fauconberg, Johanne de Gestlinge, Osbart filio Hervey, Walter de Crisping, Justiciar, et aliis Baronibus Domini Regis."

In the year 1210, King John's necessities increasing by the intrigues of the Church of Rome with his ecclesiastical subjects, he announced a parliament to meet him at S. Bride's or at his palace in S. Bride's parish, London, where he exacted of the clergy and religious persons a sum of £100,000, and £40,000 in particular from the White Friars or monks.

Soon after, the city, together with all other parts of the kingdom, were, by the Bishops of London, Ely, and Worcester, interdicted by the insolence and order of Pope Innocent, as the King would not obey his unjust and imperious command. Whereupon all churches and churchyards were shut up; divine service ceased in all places; there was no administration of sacraments except to infants and dying persons; and all ecclesiastical rites being omitted, the bodies of the dead were buried in the highways and ditches without the performance of funeral service.

Stow, who wrote about 1655, further relates that the palace of Bridewell in "aftertimes" was not used, but fell into ruins, insomuch that the very platform thereof remained for the most part waste and as it were but a lay stall for filth and rubbish,— only a fair well remained there.[1]

---

[1] When that part of the Royal Hotel where the large dining-room is situate was being built, Mr. Gruning, the architect, discovered two of the old bastions of the ancient Bridewell Palace.

A great part of this house, on the western side, was given to the Bishop of Salisbury, and hence Salisbury Square; the other remaining waste until the time of King Henry VIII.

The Holy Well of S. Bride has, with others in and about London, declined; but, inasmuch as it gave its name to the adjoining palace, hospital, prison, and at last to almost every house of correction throughout the kingdom, until quite recently, its fame may be said to be perpetuated in the fact that the whole property is termed to this day, "The Township of Bridewell," or Bridewell Precincts.

Hone's "Everyday Book" asserts that the last public use of the water of S. Bride's Well drained it so much that the inhabitants of the parish could not get their usual supply.

The occasion was a sudden demand. Several men were employed in filling thousands of bottles a day or two before the 19th July 1821, on which day His Majesty George IV. was crowned at Westminster, and Mr. Walker, of the Hotel, 10 Bridge Street, obtained it by the only means now remaining, from the cast-iron pump over the well in Bride Lane.

After various vicissitudes, Bridewell Palace and Precincts appear to have passed into the possession of Cardinal Wolsey, but on the downfall of the great prelate and statesman again reverted to the crown.[1]

It has been asserted that Wolsey copied the example of King John, and mulcted the heads of religious houses in England in enormous sums, but this is too much like the story related on page 4 to obtain a ready credence.

Cavendish says in his life of Cardinal Wolsey: "He found means to be made one of the King's Council, and to grow in good estimation and favour with the King, to whom the King gave a house at Bridewell in Fleet Street, sometime Sir Richard Empson's, where he kept house for his family, and daily attended upon the King in the Court."

In 1522 Henry VIII. repaired in the small space of six weeks, the old palace which had been much neglected, for the reception of the Emperor Charles V., and that at considerable expense. Charles V., however, lodged in the Black Friars and his suite in the palace, a gallery of communication being flung over the Fleet estuary, and a passage cut through the city wall into the Emperor's apartments.

After Wolsey's fall, Henry VIII. resided at Bridewell, particularly in 1529, and during the agitation at Blackfriars, concerning the monarch's marriage with Queen Catherine of Aragon.

---

[1] *Vide* SHAKESPEARE, *King Henry VIII.*

" The most convenient place that I can think of
For such receipt of learning is Blackfriars ;
There shall ye meet about this mighty matter."
—SHAKESPEARE, *Henry VIII.*, Act ii.

Shakespeare makes the whole of the third Act of this play pass in the palace of Bridewell, and this is historically correct.

" It was there," says Cavendish, " that the unhappy Catherine received Wolsey and Campeggio, having a skein of red silk about her neck, being at work with her maidens."

Hall, in his Chronicle, narrates it thus : " In 1528 Cardinal Campeius was brought to ye Kinge's presence, then living at Brydewell, by ye Cardinal of Yorke, and was caryed in a chayer of crimson velvet borne between iiii persones, for he was not able to stand, and the Cardynall of Yorke and he sat both on the ryght hand of the Kinge's throne, and there was one Francisci, Secretary to Cardinal Campeius, made an eloquent oracion in the Latin tongue. And the same King caused al his nobilitie, judges, and counsaylors, wᵗ divers other persons to come to his palace of Brydewell on Sunday the viii day of November at after none in his great chamber, and there delivered a speech to them concerning his marriage with Catherine of Aragon."

In the following year Henry and his queen resided at the Bridewell Palace while the question of their marriage was pending at the Blackfriars, after which, taking a dislike to the place, the King allowed it to fall into decay, in which state it remained until its appropriation to charitable uses in the following reign.

In 1525 a parliament was held in Blackfriars, and in Bridewell Palace the King created states of nobility, to wit:

Henry Fitzroy, a child (whom he had by Elizabeth Blunt) to be Earl of Nottingham, Duke of Richmond and of Somerset, Lieutenant-General from Trent Northward, Warden of the East, Middle, and West marshes for anenst Scotland.

Henry Courtney, Earl of Devonshire, cousin-german to the King, to be Marquis of Exeter.

Henry Brandon, a child of two years old, son of the Earl of Suffolk, to be Earl of Lincoln.

Sir Thomas Manners, Lord Rosse, to be Earl of Rutland.

Sir Henry Clifford to be Earl of Cumberland.

Sir Robert Ratcliffe to be Viscount Fitzwater.

Sir Thomas Belvin, Treasurer of the King's household, to be Viscount Rochfort.

There is a curious MS. in the possession of the Society of Antiquaries, which gives a list of the harness for horse and footmen in Bridewell Palace on the death of King Henry VIII.

It runs as follows :—

Horsemen's Harness.

BRIDWEL.	Remaining.	Wanting.
Demilaunces brests backes .	iiij.^{xx} 80	
Collars . . .	60	20
Vambraces . .	73 pairs.	7 pairs.
Cusshes .	80 ,,	
Gauntletts .	80 ,,	
Poldernes .	72 ,,	8 ,,
Headpieces . .	40 ,,	40 ,,

Fotemen's Harness.

	Remaining	
Almaynes rivetts, brests, and backs .	150	} the sallets
Splints	150 pairs.	} and gorgets

The Odd Harness Remaining for Horsemen.

Cusshes	62 pairs.
Hand hammers	48 ,,
Gauntletts .	14 ,,
Backs . . .	9 ,,
Headpieces (broken) . .	2 ,,

The Odd Harness for Footmen.

Brests . . .	127 pairs.
Splints .	207 ,,
Skulls . . .	7 ,,

All which parcells remain in the custody of Thomas Wolner, the King's armourer.

CHAPTER II.

THE FLEET.

THE stream running hard by the old palace shared in the gradual changes that came over the city suburbs as the years rolled on. Fed by the upper rivulets, it became, by the time it reached the old Bridewell Palace, a river that joined the Thames at this very spot.

Stow says that it was called the "river of wells," certainly as early as the time of Edward I. The Conqueror, in his Charter to the College of S. Martin le Grand, had given as a boundary "from the North Corner of the Wall by the portion of Cripplegate (as the river of the bed then neere running departeth the same moore from the wall) into the running water which entereth the city."

At a parliament held at Carlisle in 1307, a complaint was laid before King Edward I., by Henry Lacy, Earl of Lincoln, that "whereas (in times past) the course of water running at London under Old-bourne Bridge and Fleet Bridge into the Thames had

been of such bredth and depth that 10 or 12 ships,
navies at once with merchandises were wont to
come to the foresaid bridge of Fleet, and some of
them unto Oldbourne Bridge, now the same course
(by filth of the Tanners and such others) was sore
decayed, also by raising of wharves but especially by
a diversion of the water, made by them of the New
Temple, for their milles standing without Baynard's
Castle in the first year of King John, and by others
divers impediments, so that the said ships could
not enter as they were wont and as they ought.
Wherefore he desired that the Maior of London, with
the Sheriffs, and other discreet Aldermen, might be
appointed to view the said course of the said
water, and that, by the oaths of good men, all the
aforesaid hindrances might be removed, and is to be
made as it was wont of old.

"Whereupon Roger le Brabason, the Constable of
the Tower, with the Maior and Sheriffs, were assigned
to take with them honest and discreet men, and to
make diligent search and enquiry how the said river
was in former time, and that they leave nothing that
may have hurt or stoppe it, but keep it in the same
estate that it was wont to be."

The river was cleansed, the mills removed, and
other things done for the preservation of the course,
but never to its old breadth and depth; the name
"river" ceased, and it was termed a "brook," *e.g.*,

Turnemill or Treemill Brooke, and this, obviously, from the number of mills erected on it.

It was "cleansed" very often, but, last of all, to any effect, in the year 1502, the 17th of Henry VII., when the whole course of "Fleet Dike" was scoured down to the Thames; and boats, with fish and fuel, were rowed to the Fleet Bridge and Oldbourne Bridge,

THE OLD FLEET BRIDGE.

as of old time, and to the great accommodation of the citizens.

In 1589 another cleansing and scouring was enforced; but, although much money was collected for the purpose, the result was only failure, and in a few years it became more cloyed than ever it had been before.

Pennant writes, that over the tidal Fleet were four stone bridges, and on the banks extensive quays and warehouses; and so great was its utility, that in 1606 no less a sum than £28,000 was spent in keeping the channel clear.

Over the Fleet without Ludgate, was a bridge of stone coped with iron pikes on either side; on the south side were placed lanthorns, which were lighted in winter evenings for commodity of travellers.

Stow writes of it :—"This bridge hath been farre greater in times past, but lessened as the water-course hath been narrowed; next this there is a breach in the walls of the city" [the city wall was on the eastern side of the Fleet, now New Bridge Street], "and a bridge of timber over the Fleet Dyke betwixt the Fleet Bridge and the Thames, directly over against the House of Bridewell."

In 1608 twelve large granaries were erected in the Hospital precincts at the expense of the City (capable of holding 6000 quarters of corn), and two houses for coals. These storehouses for coal in James the First's time held 4000 loads of coal: a certain Alderman Leman took great care and pains in the contriving this useful work.

Among the records preserved among the archives of the City of London, A.D. 1579–1664, there are one or two letters referring to this interesting matter.

One from the Lords of the Council to the Lord

Mayor, requiring him to see that the several Companies speedily provided their full proportion of corn, and expressing their belief that he would act therein as might be most desirable, according to what had been so worthily performed by his predecessors, by whose care had of late been built fair and large granaries for storage at Bridewell. The Council further required him to take measures that neither regrators nor forestallers enhanced the markets and thereby raised the price of corn.

This letter is dated 21st January 1612, and refers to a preceding letter (7th January 1612), acquainting the Lord Mayor that a petition having been presented to the Lords of the Council by the Company of Eastland Merchants for bringing in of corn from abroad free of custom, they had given orders accordingly, with this addition, that if by reason of plenty such corn could not be sold by them at remunerative rates, they might transport it elsewhere within the kingdom, or into foreign parts, free of custom. *Special* order should be taken that the granaries at the Bridge House and Bridewell should be ready for the stowage of corn.

Another, twelve years later, 14th January 1624, from the Lords of the Council to the Lord Mayor and Aldermen, requests them to give directions for the delivery to Sir Allen Apsley, one of the surveyors-general for victualling the navy, of 2000 quarters of wheat from the storehouses at the Bridge House, Bridewell, and elsewhere, to be made into biscuit with all expedition, to be repaid by him so soon as he could purchase it, he in the meantime leaving such a sum of money in the hands of the treasurer of the subsidies as the wheat should be indifferently praised at. And further requesting them to permit the said surveyor to use the granaries, bakehouses, and cellars at the Bridge House and Bridewell, as he might require for His Majesty's service, between then and Midsummer following.

By an Act of Parliament passed in 1756 the magistrates of the City of London were empowered to erect a stone bridge across the river Thames from Blackfriars to the opposite shore in the county of Surrey. They were also authorised to fill up the channel of the Fleet Ditch, and to purchase and remove such buildings, the removal of which might be thought proper for forming and widening streets and avenues.

The limpid stream had become a public nuisance and scandal; it formed a subject for the lampoons and literary satires of the day.

Pope, in the "Dunciad," ii., thus alludes to it :—

> " This labour past, by Bridewell all descend,
> As morning prayer and flagellations end,
> To where Fleet Ditch with disemboguing stream
> Rolls the large tribute of dead dogs to Thames,
> The King of Dykes ! than whom no sluice of mud
> With deeper sable blots the silver flood."

Swift's trenchant but unsavoury lines in his "City
Shower" showed it was time to fill up the Fleet
Ditch.

Strype enters at length into the Act for making
the Fleet Ditch navigable.

FLEET DITCH, 1749.

The cost was to be borne and first defrayed "out
of the fourth part of the imposition to be raised
on coals; and in the next place, after the charge
before mentioned shall be borne out of the said
fourth part, satisfaction shall be made out of the

said fourth part to the proprietors whose ground shall be laid open, or from whom ground shall be taken;" in fact, composition was to be made.

The work of making the channel navigable, begun in 1668, lasted five years, and was finished in 1673.

The distance is 2100 feet from the Thames to Holborn Bridge, but the work reached only to *Fleet Bridge*, the rest being arched over, and covered with a new market. "It is wharfed on both sides with stone and brick laid with terras. It hath a strong campshot all along on both sides, over the brick wharfing, with land ties in several places. It hath rails of oak, breast-high above the campshot, to prevent danger at night.

"There is 5 feet water at Holborn, and that at a five o'clock tide, which is a slack one, but more at spring and other nepe tides. The wharfs are 35 feet broad, with fair buildings. The charge of sinking, clearing, wharfing, &c., £27,777, beside what was paid as composition."

In the "Remembrancia" there is preserved a letter from William Herbert (third Earl of Pembroke) to the Lord Mayor and Aldermen, dated Amesbury, 15th July 1624, requesting them to give directions for the necessary repair of the City wall near Bridewell, which ran along his house and garden in Blackfriars, and which was marvellously broken down and

B

decayed through the daily resort of barges into that dock.

The two bridges, one at Fleet Lane and the other at Bridewell, stood on stone arches over the river, having steps to ascend and descend on the other side, and half a pace over the arches, all of Purbeck and Portland stone.

So comes it that the bridge now existing at Blackfriars fills the mouth of the once filthy Fleet Ditch or Bridewell Dock. This bridge was originally named Pitt's Bridge, in honour of William Pitt, Earl of Chatham.

Sir Thomas Chitty, Knight, Lord Mayor, laid the first stone, October 30, 1760. Mr. Mylne designed and built the structure, which consisted of nine arches, that in the centre being 100 feet in span; the whole length, 995 feet, with a carriage-way of 28 feet, and side-walks 7 feet each.

It was completed at the end of the year 1768; the total cost being £152,840, 3s. 10d., defrayed by certain tolls; upon the credit of which the City magnates were directed to raise a sum not exceeding £30,000 in one year till £160,000, the given limit, was raised.

A hundred years later a new and finer bridge was erected by Cubitt, and is called Blackfriars Bridge, forming a fitting termination to the Thames Embankment.

In 1812 the following was the number of persons, horses, and carriages, &c., passing over the bridge in one day :—

Foot-passengers	61,069
Horses	822
Coaches	890
Waggons	533
Gigs, &c.	590
Carts, &c.	1,502

This presents something of a contrast with the figures which represent the traffic of the present day, represented in part by the following figures :—

On the 30th April 1888, between the hours of 8 A.M. and 6 P.M., there passed over the bridge some 6155 vehicles going north, and 6090 going south.

Considering the number of these drawn by two horses, and heavy waggons with three or more, we may fairly take an average of say two and a half horses for each, which gives 30,612 as the number of horses passing in 1888. And it is to be remembered that in 1812 there were only three bridges over the Thames for the City of London ; in 1888 there are thirteen, besides five railway bridges.

There are a few curious reminiscences of Bridewell Dock in some well-known tradesmen's tokens of the seventeenth century.

It was formerly a landing-place for Thames watermen, and, as might be expected, abounded with houses of entertainment.

In Lodowick Barry's "Ram Alley, or Merrie Tricks" (printed in 1611, 4to), Will Smallbankes and the rest of his fellows, while being conducted after supper by torchlight from the Mitre in Fleet Street to the Savoy, are set on, swords drawn, by Throat and his desperadoes, who carry off the pretended heiress unperceived towards St. Giles. Thos. Smallbankes, nettled at this ill-luck, affirms that she had run off towards Fleet Bridge; but Will asserting it as a thing not possible, Thomas reiterates—

> "Upon my life,
> They went in by the Greyhound, and so strooke
> Into Bridewell,—to take water at the dock."

The "Greyhound" was a well-known tavern on the south side of Fleet Street.

The following are the tokens in Beaufoy's list connected with Bridewell :—

Bridewell Dock.

No. 230. At ye Pyd Bull In. A bull in the field. *Rev.:* Ould Brid-well 1652. In the field, M. A. E.

No. 231. Robert Chapman at Bride, Woodmonger's Arms. *Rev.:* Well-Dock. His halfpenny. In the field, R. E. C.

No. 233. Gile Ray, Woodmong. Woodmonger's Arms in the field. *Rev..* At Bridewell Dock. In the field, G. I. R.

Bridewell Steps.

No. 234. At the Sun Tavern. The sun in rays in the
field. *Rev.:* Upon Bridewell Steps. In the
field, A. E. C.

Bride Lane, Fleet Street.

No. 239. Will Hearne at ye Whit. A bear in the field.[1]
Rev.: In Bride Lane. In field, W. M. H.

---

[1] As early as 1252 a white bear was sent from Norway as a present to King Henry III. It was kept at the Tower, and fourpence
a day for his keep was directed to be paid by the Sheriffs of London. A white bear, with collar and chain and muzzled, was the
badge of Anne, consort of Richard III. The bear was the badge
of the Earl of Warwick, who was supposed to have derived it from
Urso d'Abilot.

CHAPTER III.

THE FOUNDATION (Temp. Ed. VI.)

KING HENRY VIII., upon the petition of Sir John Gresham, granted charters of Bethlehem[1] and St. Bartholomew Hospitals to the City a very few days before his death in January 1547. The circumstances which led to this are full of interest.

After the suppression of the monasteries and other religious houses, London became filled with multitudes of dissolute and necessitous persons, who before that period had depended on ecclesiastical charity for their support. It therefore became necessary to adopt some plan for the correction of offenders, and to afford a refuge and relief to such as were in actual want.

The first effort towards this laudable and charitable end was made by the pious Bishop of London, Nicholas Ridley, in the reign of Edward VI. So interesting is the account of Bridewell in connection with this in Strype's "Stow," vol. i. p. 197, that it is transcribed here.

---

[1] In the charter the name is throughout spelt "Bethlem."

From an Oil Painting by A. VAN DER WERFF.

"There were also in the city many others of the poor necessitous sort that had neither house nor harbour to put their heads in, but were fain to lie abroad in the open streets, and divers families of other poor fain to lie under one roof.

"This did closely affect many good citizens, and particularly Ridley, the good Bishop of London, who by some means was informed of it, and moved in it. And considering that Bridewell, an old decayed house of the King's situate in the city, being very large and capacious, might be extremely serviceable to this charitable purpose, he endeavoured to find a way to beg it of the King, especially at this time, when one was about buying it of the King to put it down and convert it to his own use. And for the compassing of this, in the month of May, this charitable year 1552, he wrote a very pathetical letter to Sir William Cecyl, Knight, the King's secretary, whom he knew to be of a pious disposition, as well as much about the King, having promised the citizens to move him in the matter, because he took him for one, as he told him in his letter, that feareth God."

His moving letter ran to this tenor :—

"Good Master Cecyl,—I must be a suitor with you in our Master Christ's cause. I beseech you to be good unto Him. The matter is, Sir, that he hath been too, too long abroad, without lodging, in the

streets of London, both hungry, naked, and cold.
Now thanks be to Almighty God, the citizens are
willing to refresh Him, and to give Him meat, drink,
clothing, and firing. But alas! they lack lodging for
Him, for in some one house, I dare say, they are fain
to lodge three families under one roof. Sir, there is
a wide large empty house of the King's Majesty,
called Bridewell, which would wonderfully well serve
to lodge Christ in, if He might find friends at court
to procure in His cause.

"Surely I have such a good opinion in the King's
Majesty, that if Christ had such faithful and hearty
friends that would heartily speak for Him, He should
undoubtedly speed at the King's Majesty's hands.

"I have promised my brethren the citizens to move
you, because I do take you for one that feareth God,
and would that Christ should be no more abroad in
the street."

He prayed him also for God's sake that he would
stop the sale of the house, in case any were about
buying it, as he had heard there was, and that he
would speak in our Master's cause.

The said Bishop wrote also to Sir John Gates,
another great man at court, about the business more
at large; and he joined, he said, Cecyl with him and
all others that loved and looked for Christ's final
benediction on the latter day: meaning that in the

Gospel, "Come, ye blessed of My Father, inherit the kingdom prepared for you from the beginning of the world : for I was an hungered," &c., &c.

He also sent instructions by the bearer of this letter to confer further with Cecyl in this affair; so that Ridley's name must not be forgotten as a great instrument to the procurement of Bridewell to the City.

"Afterwards, this house being obtained to the City, it was employed for the correction and punishment of idle vagrant people and dissolute, and for setting them to work, that they might in an honest way take pains to get their own livelihood."

"And here I cannot omit to leave upon record, to their eternal honour, the names of the two good Mayors of London, Dobbs and Barnes, the former a main instrument of procuring the foundation of this and the other hospitals, the latter of furthering the good estate of them ; whom in this most Christianly affectionate manner the aforesaid good Bishop of London accosted in one of the letters he wrote out of prison a little before his death."

The following is a copy of the letter here referred to :—

"O Dobbs, Dobbs, Alderman and Knight, thou in thy year didst win my heart for evermore for that honourable act, that most blessed work of God, of the erection and setting up of Christ's holy hospi-

tals and truly religious houses, which by thee and
through thee were begun.   For there, like a man of
God, when the matter was moved for Christ's poor
silly members to be holpen from extreme misery,
hunger, and famine, thy heart, I say, was moved with
pity ; and as Christ's high honourable officer in that
cause, thou calledst together thy brethren the Aldermen of the City, before whom thou breakest the matter
for the poor.   Thou didst plead their cause ; yea, and
not only in thine own person thou didst set forth
Christ's cause, but to further the matter, thou brought-
est me into the Council Chamber of the City, before
the Aldermen alone, whom thou hadst assembled there
together to hear me speak what I would say as an ad-
vocate by office and duty in the poor man's cause.

"The Lord wrought with thee, and gave thee the
consent of the brethren, whereby the matter was
brought to the Common Council, and so to the whole
body of the City, by whom, with an uniform consent,
it was committed to be drawn, ordered, and devised
by a certain number of the most witty citizens and
politic, endued also with godliness and with ready
hearts to set forward such a noble act, as could be
chosen in all the whole City.

"And they, like true and faithful ministers both to
the City and their master Christ, so ordered, devised,
and brought forth the matter, that thousands of poor
silly members of Christ, that also for extreme hunger

and misery should have famished and perished, shall be relieved, holpen, and brought up, and shall have cause to bless the Aldermen of that time, the Common Council, and the whole body of the City, and especially thee, O Dobbs, and those chosen men by whom this honourable work of God was begun and wrought.

"And thou, O Sir George Barnes, thou wast in thy year not only a furtherer and continuer of that which before thee by thy predecessor was well begun, but also thou didst labour so to have perfected the work that it should have been an absolute thing and a perfect spectacle of true charity and godliness unto all Christendom.

"Thine endeavour was to have set up an house of occupation, both that all kind of poverty, being able to work, should not have lacked whereupon profitably they might have been occupied, to their own relief and to the profit and commodity of the commonwealth of the City, and also to have retired thither the poor babes brought up in the hospitals, when they had come to a certain age and strength, and also all those which in these hospitals aforesaid have been cured of their diseases.

"And to have brought this to pass thou obtainedst, not without great diligence and labour, both of thee and thy brethren, of that godly King Edward, that Christian and peerless prince, the princely palace of

Bridewell, and what other things to the performance
of the same, and under what conditions, it is not
unknown.

"That this thine endeavour hath not had like success, the fault is not in thee, but in the condition and
state of the time."

By the contents of which letter, may be understood
more particularly than perhaps any history hath yet
told us, what was that course and method which the
citizens took, in their first attempts in founding
Bridewell and Christ's Hospitals.

But while these before-mentioned good motions
were in hand in the City, the King was excited to
these charities by good sermons preached before
him.

Such was that of Mr. Lever, a learned and pious
preacher in those days, and Master of St. John's
College, Cambridge, who, in a Lent sermon before
His Majesty, used these words :—

"O merciful Lord, what a number of poor, feeble,
halt, blind, lame, sickly, yea, with idle vagabonds
and dissembling caitiffs mixed among them, lie and
creep, begging in the miry streets of London and
Westminster? It is too great pity afore the world,
and to utter damnation afore God, to see these
begging as they use to do in the streets.

"For there is never a one of these but he lacketh

either thy charitable alms to relieve his need, or
else thy due correction to punish his fault.

"These silly souls have been neglected throughout
all England, and especially in London and Westminster; but now I trust that a good overseer, a
godly bishop—Bishop Ridley, I mean—will see that
they in these two cities shall have their needs relieved
and their faults corrected, to the good ensample of all
other towns and cities.

"Take heed that there be such grass to sit down
there, as ye (speaking to the King) command the
people to sit down; that there be sufficient housing
and other provision for the people there, as ye command them to be quiet. 'The men sat down above
5000 in number,'" which words were part of the
Gospel for the day, out of which he took his text.

And Ridley, that zealous and charitable prelate,
and a true father of his flock in London, was season-
ably called also to preach before the King at Westminster, when he so closely and affectionately pressed
persons in high place and calling to be instruments in
helping and succouring the poor, that the King was
exceedingly moved with his discourse, and presently
sent for him, taking notice to him of his sermon,
and that he supposed he had him chiefly in his
eye, as being the highest of those in great place and
calling, that he the Bishop spake to.

Then the King assured him of his own readiness

to promote such good purposes, desiring him to
direct him therein, and what he would advise him
to do in that part.

How the Bishop hereupon referred the King to
the City, and how the King presently caused a letter
to be penned and sent to the Mayor and his brethren
to enter into consideration about it, and how the
Mayor, the Bishop, and other eminent citizens met
together to prepare something for the King on behalf
of their poor, and the report thereof made to the
King, and other matters relating hereunto, are all set
down under "Christ Church" in the Ward of Farringdon Within as follows :—

"Doctor Ridley, then Bishop of London, came
and preached before the King's Majesty at Westminster, in which sermon he made a fruitful and
godly exhortation to the rich, to be merciful unto
the poor, and also to move such as were in authority
to travel by some charitable ways and means to
comfort and relieve them.

"Whereupon the King's Majesty, being a prince
of such towardness and virtue for his years as England before never brought forth, and being also
so well retained and brought up in all godly knowledge, as well by his dear uncle the late Protector
(Edward Seymour, Duke of Somerset), as also by
his virtuous schoolmasters, was so careful of the
good government of the realm, and chiefly to do

and prefer such things as most especially touched the honour of Almighty God; and understanding that a great number of poor people did swarm in this realm, and chiefly in the City of London, and that no good order was taken for them, did suddenly and of himself send to the said Bishop as soon as his sermon was ended, willing not to depart until that he had spoken with him.

"And this that I now write was the very report of the said Bishop Ridley, who, according to the King's command, gave his attendance.

"As soon as the King's Majesty was at leisure, he called for him, and caused him to come unto him in a great gallery at Westminster, where to his knowledge, and the King likewise told him so, there were present no more persons than they two; and therefore made him sit down in one chair, and he himself in another, which, as it seemed, were before the coming of the Bishop there purposely set, and caused the Bishop, in spite of his teeth, to be covered, and then entered communication with him in this manner.

"First, giving him hearty thanks for his sermon and good exhortation, he therein rehearsed such special things as he had noted, and that so many that the Bishop said, 'Truly, truly'—for that commonly was his oath—'I could never have thought that excellency to have been in His Grace, but that I beheld and heard it in him.'

"At the last, the King's Majesty much commended him for his exhortation for the relief of the poor.

"The Bishop, thinking least of that matter, and being amazed to hear the wisdom and earnest zeal of the King, was, as he said himself, so astonished, that he could not tell what to say; but after some pause said that he thought at this present for some entrance to be had, it were good to practise with the City of London, because the numbers of the poor there 'are very great, and the citizens also are many and wise,' and he doubted not but that they were also both pitiful and merciful, as the Mayor and his brethren, and other the worshipful of the said City; and that if it would please the King's Majesty to direct his gracious letters unto the Mayor of London, inviting him to call in such assistants as he should think meet to consult of this matter for some order to be taken therein, he doubted not but good would follow thereon.

"And he himself promised the King to be one himself that should earnestly assist therein.

"The King forthwith not only granted his letter, but made the Bishop tarry until the same was written, and his hand and signet set thereto, and commanded the Bishop not only to deliver the said letter himself, but also to signify unto the Mayor that it was the King's especial request and express commandment that the Mayor should assist therein, and, as soon as he might conveniently, give him knowledge how far he proceeded therein.

"The Bishop was so joyous of receiving of this letter, and that now he had occasion to assist in so good a matter, wherein he was marvellous zealous, that nothing could have more pleased and delighted him.

"Wherefore the same night he came to the Mayor of London, who was then Sir Richard Dobbs, Knt., and delivered the King's letter, and showed his message with effect.

"The Lord Mayor not only joyously received this letter, but with all speed agreed to set forward the matter, for he also favoured it very much. And the next day being Monday, he desired the Bishop of London to dine with him, and against that time the Mayor promised to send for such men as he thought meetest to talk of this matter; and so he did. (*See* p. 35.)

"The order taken, and the citizens meeting to further the same, the report was submitted, and His Grace (the King), for the advancement thereof, was not only willing to grant to such as should be overseers and governors of the said houses [1] a charter of corporation and authority for the government thereof, but also requested that he might be accounted as the chief Founder and Patron thereof.

---

[1] Christ's, Bridewell, and S. Thomas.

C

" And for the furtherance of the work and the maintenance of the same, he of his mere mercy and goodness granted that whereas before, certain lands were given to the maintaining of the house of the Savoy founded by King Henry VII. for the lodging of pilgrims and strangers, and that the same was now made but a lodging for loiterers, vagabonds, and dissolute, that lay all day in the fields, and at night were harboured there, which was rather the maintenance of beggary than any relief to the poor, he gave the same lands, being first surrendered by the Master and Fellows, which lands were of the yearly value of £600, unto the City of London for the maintenance of the aforesaid foundation.

" And for a further relief, a petition being made for a license to take in mortmain or otherwise without license, lands to a certain yearly value, and a space left in the patent for His Grace to fill in the sum it might please him, he with his own hand wrote the sum 4000 marks by the year, and then in the hearing of the Council he uttered this prayer, 'Lord, I yield Thee most hearty thanks that Thou hast given me life thus long to finish this work, to the glory of Thy name.'"

He died two days afterwards, having by this, almost the last act of his life, become the founder of the Royal Hospitals in London.

Perpetuating the memory of the young monarch, his picture was to be found in Strype's time hanging close to the pulpit in the chapel at Bridewell, with these lines under it :—

> "This Edward of fair memory the Sixth,
> In whom with greatness, goodness was commixt,
> Gave this Bridewell, a palace in old time,
> For a chastising house of vagrant crime."

The action taken by the Lord Mayor, Sir R. Dobbs, on receiving the Bishop's communication, was of the promptest sort. He sent for his colleagues, and made a selection of twenty-four Aldermen and commoners to be a committee. After several meetings they agreed upon a scheme.

In this they considered nine especial kinds and sorts of poor people, and those they brought into these degrees :—

Three Degrees of Poor.

1. Poor by impotency.
2. Poor by casualty.
3. Thriftless poor.

1. The poor by impotency are also divided thus :—
   (1.) Fatherless poor man's child.
   (2.) The aged, blind, and lame.
   (3.) The diseased person by leprosy, dropsy, &c.

2. The poor by casualty, thus :—
   (1.) The wounded soldiers.
   (2.) The decayed householder.
   (3.) The visited with any grievous disease.

3. The thriftless poor, thus :—
   (1.) The rioter, that consumeth all.
   (2.) The vagabond, that will abide in no place.
   (3.) The idle person, as dissolute women and
        others.

For these sorts three several houses were provided :—

First, For the innocent and fatherless, the house
    that was the late Grey Friars in London, but
    now called by the name of Christ's Hospital.
Secondly, Were provided the Hospital of St. Thomas
    in Southwark and St. Bartholomew in West
    Smithfield.
Thirdly, Bridewell, for the vagabond, idle, and
    dissolute.

They also provided outdoor relief for the decayed
householder, and pensions for the leper.

The Charter granted by the King, sets forth the
particulars of the royal gift and the various goods,
chattels, furniture, linen, &c., from the Savoy. "Utensils, beddings, and necessaries lately belonging and
appertaining unto our said late Hospital of the Savoy,"
(reserving "one great bell, and one small bell, and
one chalice,") and revenues unto the yearly value of
4000 marks, unto the said Mayor and commonalty
and citizens of the city aforesaid, and to their succes-
sors for ever."

" In strains melodious, the trump of Fame,
  Enamour'd, echoes the Sixth Edward's name ;
  A name which charity has firm impressed
  On the warm feelings of the grateful breast."

In the "Memorials of the Savoy," by Loftie, is the following passage :—

"Sir Roger Cholmely, Lord Chief Baron, was appointed visitor to report upon the state of the hospital and its revenues ; this he did, and his report was unfavourable."

Ralph Jackson was appointed Vice-Master, 9th June 1553, but had to surrender the hospital to the King, who, on the 26th June, made over the estates, with the implements and utensils, to Bridewell.

In 1558 the Hospital of the Savoy was refounded ; and it is said that all the beds having been taken away, the "ladies of the court," for "the better attaining the Queen's good grace," refurnished them in a very ample manner.

The charter of Bridewell was confirmed 26th June 1553, and in 1555, during the February of that year, Sir W. Gerard, Mayor, and the Aldermen, entered Bridewell and took possession thereof, according to the gift of the said King Edward VI.,—the same being confirmed by Queen Mary.

The following is an abstract of an Act of Common Council, the last of February, the second and third of Philip and Mary, concerning Bridewell :—

"Forasmuch as King Ed. VI. had given his house
of Bridewell unto the City, partly for the setting of
idle and lewd people to work, and partly for the
lodging and harbouring of the poor, sicke, and
weake, and sore people of the city, and of poor way-
faring people repairing to the same; and had for
this last purpose given the bedding and furniture
of the Savoy to that purpose: Therefore, in consideration that very great charges would be required
to the fitting of the said house and the buying of
tooles and bedding, the money was ordered to be
gotten up among the rich people of the Companies
of London."

CHAPTER IV.

PARTICULARS AND USES OF THE HOSPITAL AS FOUNDED.

THE education of destitute children and the cure of paupers afflicted with disease having been provided for by the erection of the Hospitals of Christ Church, St. Bartholomew, and St. Thomas, the citizens turned their attention to the establishment of a place for the reception of those vagrants and mendicants who were not the objects of these institutions, and with this view this petition (referred to, p. 33) was presented to the King, Edward VI., praying for a grant of the ancient palace of Bridewell, which appeared suitable for their purpose.

"*A Supplication made by the Assent of the Governors of the Poor in the name of the same Poor, to the King's Majesty for the obtaining of the House of Bridewell.* A.D. 1552.

"FOR JESU CHRIST'S sake, right dear and most dread Sovereign Lord, we, the humble, miserable, sore sick, and friendless people, beseech your gracious Majesty to cast upon us your eyes of mercy and compassion, who now by the operation of the Almighty God, the citizens of London have already so lovingly and tenderly looked upon, that they have not only provided help for the maladies and diseases and the virtuous education and bringing up of our miserable and poor children, but also have in a readiness, most profitable and wholesome occupation for the continuing of us and ours in godly exercise, by reason whereof we shall no more fall into that puddle of idleness, which was the mother and leader of us into beggary and all mischief, but from henceforth shall walk in that fresh field of exercise which is the guider and begetter of all wealth, virtue, and honesty.

"But also, most gracious Lord, except we find favour in the eyes of your Majesty, all this their travail and our hope of deliverance from that wretched and vile state cannot be attained, for lack of harbour and lodging, and therefore, most gracious Sovereign, hear us, speaking in Christ's name and for Christ's sake, have compassion on us, that we may lie no longer in the street for lack of harbour, and that our old sore of idleness may no longer vex us, nor grieve the commonweal. Our suit, most dear Sovereign, is for one of your Grace's houses called Bridewell, a thing no doubt both unmeet for us to ask of your Majesty and also to enjoy, if we asked the same for our simple

living and unworthiness-sake, but we, as the poor
members of our Saviour Jesu Christ, sent by Him,
most humbly sue to your Majesty in our said Master's
name, Jesu Christ, that we for His sake, and for the
service that He hath done to your Grace and all the
faithful commons of your realm, in spending His most
dear and precious blood for you and us, may receive
in reward at your Majesty's hands, given to us, His
members (which of His great mercy He accounteth
and accepteth in our behalves, as granted and given
to Himself), the same, your Grace's house as a most
acceptable gift and great obligation offered unto Him,
and then, not we but He, our said Master and Saviour, which already hath crowned your Majesty with
an earthly crown, shall, according to His promise,
crown your Grace with an everlasting diadem, and
place you in the palace of eternal glory, and not we
only, but the whole congregation and Church spread
throughout the whole world, shall and will night and
day, call and cry incessantly unto our said loving and
sweet Saviour and Master to preserve and defend
your Majesty both now and for ever."

It was ordered that the Bishop of London should
go with the deputation, amongst whom were—

Sir Martin Bowes, goldsmith.
Sir Rowland Hill, mercer.
Sir Andrew Ind, skinner.

Sir John Gresham, mercer.

Sir John Ayleph.

Master Chester, and others.

The Bishop delivered the supplication with his own hands unto the King's Highness, in his inner closet, kneeling on his knees, and then made a long and learned oration to the commendation of the citizens in the travail of their good work, and greatly stirred by wonderful persuasion the King's Majesty to be patron and founder thereof, and to further all their suits.[1]

The King's consent having been obtained, a statement was presented to the King's Majesty's most Honourable Council, A.D. 1552, setting forth the objects of the Institution.

Heartily thanking the King and the Council for their sympathy, they proceed to name how the same shall be used and how managed.

That beggary and thieving abounded; how many statutes had been made for the redress of the same, and little amendment followed; the conclusion that idleness was the cause of all this misery; therefore work was recommended, that beggary might be reformed.

That general provision of work should be made for

---

[1] The last chapter, iii., p. 22, relates more particularly Ridley's communications with the King.

the willing poor, as well as for the strong and sturdy vagabond.

That the work had already been begun to succour the indigent child, the sick and the impotent. And for the idle, that a house of occupation should be erected, wherein the child, "when he is grown up and found unapt to learning and unable to take service, may be exercised and occupied, the sturdy and idle set to work, and prisoners who are quit at the sessions." And as room is required for these purposes, the House of Bridewell is requested.

And for the use of the house, it is set forth that profitable occupations be sought, such as making of caps more substantial than those made in France.

That those lame of leg but whole of hand, should be occupied in making feather bed-ticks, wool cards, drawing of iron, spinning, carding, knitting, and winding of silk, &c. ; and that the stubborn and fouler sort be set to making of nails and other iron-work.

That as the citizens had given large sums for the furnishing of Christ's Hospital and St. Thomas', so they would do the needful for the third house.

That the whole charge be committed to thirty governors, of whom six are to be Aldermen, none to receive fee or reward, and one of their number to be appointed Treasurer for a year.

Taskmasters and mistresses to be appointed at

convenient stipends, and other officers, as steward, porters, cook, &c.

That a perfect confidence in the munificence of those desirous to do good would enable the continuance of the house to be maintained in the future.

That if the House of Bridewell be not thought proper and meet to be converted for the object in view, then to sue for the house and lands of the Savoy, &c., &c.

The indenture of covenant was made, and a charter granted to the Royal Hospitals, which purports to confer those powers of police which were essential to give effect to the intention and objects of the citizens, and to enable them to clear the city of the vagrants and mendicants by whom it was infested.

The union of Bridewell with Bethlehem was made at a court held at Christ's Hospital on 27th September 1557. Although the objects of the two hospitals were and are essentially different, the same governors acted, and do still act, for the two hospitals, as a matter of convenience.

There is in existence a small black-letter volume, 1557, which gives the "Ordinances for the Good of the Hospitals;" among other things, directions to the beadles to walk through their wards, staff in hand, two and two, and apprehend and convey to Bridewell all vagrants and idle persons; if aid be required, to call the first constable to assist ; to report if the constable does not do his duty ; to see that no rogue or idle person resort to trouble the street whenever a citizen lies dead within their walks ; to receive reward gratefully for their work ; to call the assistance of other beadles, if need require, to help to clear the streets, and to pay them accordingly ; to attend at St. Paul's Cross at sermon-time ; to apprehend all vagrants and idle persons, women and children, to Bridewell ; and if found negligent, to have their staffs taken from them and excluded for ever from serving.

The Ordinances and Rules drawn out for the governors (or rather almoners) of Bridewell are very long and wordy, but the pith and marrow of them is—the' importance of succour and relief for the poor, sick, and aged ; to yield alms to the poor and honest decayed householder ; to train up the beggar's child in virtuous exercise, and to compel the wilfully idle and dissolute to better ways, for which purposes the houses of Christ's and St. Thomas' and Bridewell were provided. For the latter house, such and such of the governors are appointed, some for the oversight of cloth-making, others to the smithy and nail-making, and some to the millhouse and bakehouse ; some to receive offenders, and examine and punish the same ; to visit taverns, alehouses, dicing-houses, bowling-alleys, tennis-plays, and all suspected places of evil resort in and about London.

And by consent of a full court, to make alterations from time to time in the rules and management
of the estates.

For the clothmaking.—Women might be employed
in spinning or carding, and the whole under the supervision of the chief workmaster, who was to correct
anything that might be unseemly.

For the nailhouse.—Apprentices were to be brought
in and taught. An inventory of all stock was to be
most carefully kept, and provision of iron and sea-coal
to be provided for the occupying of " our people."

For the millhouse and bakehouse.—To permit no
loitering of the vagabonds employed in them. To
appoint eight persons to the less mill, ten to the
greater, and each to grind daily two bushels. To
observe good order, and certain told off to keep the
place thoroughly sweet and clean, and to work at
making tile-pins if any sat at leisure.

About 250 quarters of wheat were required for the
use of the three hospitals, and the stewards of Christ's
and St. Thomas' were to be careful in their tallies for
what was issued to them from the Bridewell Mills.

A comptroller of the diet of the house to be
appointed, to whom only the care and provision of
victuals was committed. He was to buy wood and
not provide charcoal (" for that is unprofitable "). He
was to be very careful about keeping stores under lock
and key, to see that the wards and their contents be

clean, to keep a strict inventory, and make proper
provision for sheets, shirts, or any other necessary;
"and this is your charge, which God grant you to
perform and reward your pains."

All these charges are quaintly drawn up. That of
the bailiff of the Mill and Bakehouse is as follows :—

"And this your charge, wherein if ye travail diligently, and chiefly now at the first, for good order's
sake, good men will commend you, all the worthy
governours must love you, and God Almighty will
bless you here with worship, and reward you in heaven
with the crown of glory everlasting."

In the report above cited there is a record of the
Order of Common Council made 4th August 1579.

In it there are 53 orders and 13 additional precepts
for provision and money, to be had for good uses.
Among the orders it sets forth that there are to be set
up in Bridewell certain arts, occupations, works, and
labours, stock and tools for those works to be provided, also bedding, apparel, and diet. Great care to
be taken that vagrants not belonging to the city, depart
for their own places of birth or last abode, and to be
apprehended when sufficient time had elapsed after
the proclamation to this effect had been made. If
any returned in a roguish manner he was to be openly
whipped at a cart's tail and sent off with passport for
the second time. If any one "eft sones" (often)
return to be used as a rogue of the first degree, and

if offending again as a rogue in the second degree,
viz., as a felon according to the law.

Should any be sick, the Hospitals of St. Bartholomew
and St. Thomas to receive such. Those whom the
City was charged by law to provide for and able to
work, to be received into Bridewell and kept to work
on very meagre diet, and to be punished if they were
idle. If any run away or escape, and be retaken as a
vagrant, he is to be whipped at the cart's tail, and
on a repetition, to be treated as a rogue of the first
degree and have his ear bored, and at the third similar
offence, he is to be used as a felon.

Citizens may take clever vagrants, skilful in any
occupation, into their service. Children of those
chargeable to the City and unable to keep them, to
be sent to Christ's Hospital. The parish to provide
for the aged and impotent, and if these be found
begging, punishment to be administered to such in
Bridewell.

No one to leave children or other belonging to
other places in the City under pain and penalty.
Citizens to hand over beggars to the beadles under
pain of 3s. 4d., and constables to apprehend the
same, in pain of 6s. 8d. both by night and day; if
at night to convey the offender to the cage or
counter (cages were first set up in 1503 by order of
Sir William Capell, Lord Mayor, in every ward for
the punishment of rogues and vagabonds), and to

Bridewell the next day.   Aldermen to hold inquisi-
torial wardmotes under pain of 40s. as a safeguard
against vagrants, and for the better reformation of
the idle youth and unthrifty poor.

A vestry to be held a week previous to the ward-
mote "to enquire and understand of all idle persons, vagabonds, rogues, disordered parents or
masters of houses, disordered children of the poor,
disordered alehouses, and such like; absences from
church, and other misdemeanours of the parish."

No relief to be granted to the idle but able-bodied,
but work to be found for the willing unemployed.

Unruly youth, if incorrigible, to be punished at
Bridewell.

The Governors of Bridewell, by virtue of their
charter, to render all assistance possible to the
Aldermen and their deputies in the execution of their
duties, should dealing with these exceed their powers.

Two Governors to be appointed for every art,
science, or labour, as overseers, and four to attend
two hours a day for the examination and direction
of those brought to the house after the first search.

Certain fines to be levied upon those appointed
to see the works at Bridewell properly carried on,
for non-attendance, but due notice of attendance
to be given.

The Savoy to be searched as to proper persons
taking advantage of this temporary asylum and

D

refuge; and the like abuse to be seen to at St. Thomas' Hospital.

Alehouses, tippling-houses to be reformed; not too many permitted, and the number abridged to a reasonable proportion in each ward.

Only those allowed to be open which bear a good character. Bonds to be enforced for the good observance by victuallers against these orders; and where large houses are converted into smaller, and alleys farmed out, all victualling and drinking shops to be disallowed to the landlords. Punishment by whipping to be at the discretion of the Governors without waiting for the Guildhall Sessions.

The following arts and occupations, labours, and works to be set up in Bridewell :—

> The work in the mills.
> The work in the lighters and the unlading of sand.
> The carrying of sand.
> Making of gloves.
> Making of combs.
> Making of inkle and tape.
> Making of silk lace.
> Making of apparel for the house.
> Spinning of woollen yarn.
> Knitting of hose.
> Spinning of linen yarn.
> Spinning of candlewick.

Making of packthread.

Drawing of wire.

Making of pins.

Making of shoes.

Thicking of caps by hand and foot.

Making of woolcards.

Making of nails.

Making of points.

Making of knives.

Making of baize.

Making of brushes.

Making of tennis balls.

Making of felts.

Picking of wool for felts.

Or any other that may fall in practice.

Reformation, and not perpetual servitude, was the real object to be attained if possible, and every effort to be made that youths who might be sent to Bridewell should be apprenticed there to a trade, sent to service, or to sea.

Employers in various trades to be encouraged to send work to Bridewell, so as not to let the productions of the house be injurious to trade. To be careful about the foreigners, lest they become burdens to the City.

Faulty and forfeited leather to be used up solely for the poor in Christ's Hospital and Bridewell.

Then follow the rules and regulations for provision, and money to provide for diet, bedding, tools, and stock. Two-fifteenths to be assessed and levied in usual manner by the body of the City, foreigners to be taxed, and assessments to be received according to circumstances from time to time.

Playing of interludes considered degenerating to the morals of the youthful, risky by reason of concourse as regards the plague, wasteful of time, and drawing folk away from the service of God, so must be interdicted in the time of Lent and Easter, and every holiday and Sunday in the year.

Forfeiture of bonds to be strictly enforced, and money to be obtained for the maintenance of the Hospital by sermons at the Cross, and other legitimate means.

Citizens, artificers, farmers, and gentlemen to be solicited for situations for servants and children out of Christ's Hospital and Bridewell for their kitchen and service, with the offer to provide them with convenient apparel, and bind them for any competent number of years, and as a further inducement, to give them thorough instruction in reading, writing, grammar, and music.

In September 21, 1579.—Governors were appointed for the several offices.

3 to be surveyors of shoemakers.
4 for the house.
4 for land and lime, &c.

From this it is clearly evident that sand and gravel were raised from the bed of the river, but the lime-kilns were discontinued by order of Chancery in the reign of Elizabeth.

The Order of Common Council, October 11, 1587, calls attention to the advisability of clearing the City of vagrants with which it was infested, and after proclamation to cause all who had not been resident for three years past, to depart to their last place of residency.

The stout and strong vagabonds to work at scouring the town ditches, and to be fed from Bridewell; the sick to be sent to St. Thomas', and the infants and little children to be kept at Christ's.

And to the better avoiding idle beggars and vagabonds, humble suit be made to the Privy Council, that letters be written to the heads of the neighbouring counties to pass them on to their settlements.

The object of the citizens in the erection of the Royal Hospitals, and the necessity, in order to its attainment, of an entire co-operation between them, is apparent; but Bridewell was gradually separated from the other hospitals, and became a prison and a place for the reception of apprentices.

The powers of police which the Charter purported to confer, extended equally to all the hospitals, though the peculiar province assigned to Bridewell, caused the exercise of them to devolve chiefly upon the governors of that establishment, and rendered it more permanent there than in the other hospitals.

Mere immorality was thought within their cognisance, and people were frequently taken into the house on slight and insufficient grounds, and on charges which ultimately were proved to be false; very often considerable oppression and injustice occurred, inasmuch as prisoners were received upon simple complaint and without legal warrant.

The exercise of this power, and a practice that prevailed from the period of its Foundation, of sending people to Bridewell merely to receive corporal punishment, in course of time caused one part of the establishment to assume the character of a mere prison; while the trades and manufactures, already alluded to, by degrees degenerated into an expansive and useless establishment of persons called "art masters," to whom boys, in a sense quite foreign to the purpose of the charter, were bound apprentice. And these "art masters" were not got rid of until a much later period. (*Vide* chap. ix.)

Christ's Hospital, doubtless, was intended for children, who in infancy might receive a virtuous education and bringing up, while Bridewell was for "the child, when grown up," so that in full age he should not lack matter whereon he might virtuously occupy himself in some good occupation or science, profitable to the common weal.

CHAPTER V.

PENAL DISCIPLINE.

THE following are some of the records of the infliction of corporal punishment in Bridewell :—

December 1556.—A woman, resident in Southwark, was judged by the Lord Mayor and Aldermen to be whipped at Bridewell, and sent to the Governors of Christ's Hospital for a further reformation, and subsequently to be placed in the pillory in Cheapside with a paper in her hand, whereon was written—" Whipped at Bridewell for leaving and forsaking her child in the streets."

July 16, 1559.—A woman named Jane Foster was brought into the house for enchanting Margaret Storer, and trying to bring her into dissolute and evil ways.

July 19, 1563.—The order devised and taken by the Governors of Bridewell for cutting off the hair of the head of such immoral women as were committed to the said house, and would not be quietly contented to reform and amend themselves, by whipping and other punishments, was confirmed.

April 1574.—A scold named Joan Grove threatened with threescore stripes with a whip, if ever again, she be proved to exclaim with her tongue against Sir William Drury's man or any other.

April 24, 1577.—Seven persons, being common rogues, had correction and were discharged.

It would also appear (from the following extracts from Jardine) that torture was plentifully used at Bridewell in the reign of Elizabeth.

In the Council Book there is a warrant dated October 25, 1591, directing Dr. Fletcher Richard Topcliffe (the well-known instrument of Government for the discovery of recusants) and two other persons "very straightly" to examine "Eustace White," a seminary priest, and one "Brian Lulsy," a distributor of letters to papists, and if they refuse to answer directly, to put them to the manacles and such other tortures as are used in Bridewell.

This is the first occasion on which this instrument finds mention, but from this time it was by far the most usual kind of torture.

It seems to have been kept at Bridewell until about the year 1598, after which time it is mentioned in warrants as one of the tortures commonly in use at the Tower.

"I cannot discover," says Jardine, "from any credible authority, of what the 'manacles' consisted."

It is perhaps worthy of remark that at the present day a variety of instruments of torture are shown at the Tower, and visitors are assured that they were taken from the Spanish Armada in 1598, the exact date at which the manacles were introduced at Bridewell.

One of the instruments now at the Tower, which compressed the neck of the sufferer down towards his feet, might be the "manacles," and if so, Shakespeare probably alludes to it when he makes Prospero say in the "Tempest"—

> "He is a traitor!
> I'll manacle thy neck and feet together."

At the Tower, however, this instrument of torture is called the "scavenger's daughter."

October 29, 1591.—A warrant issued to Attorney and Solicitor General (Popham and Egerton) to examine Thomas Clinton, a prisoner in the Fleet, and if he does not deal plainly in his answers, to remove him to Bridewell, "there to be put to the manacles and such torture as is there used."

June 4, 1592.—Warrant issued. Owen Edmondes to the torture in Bridewell.

February 8, 1593.—Austin, Bagshaw, Ashe, to be removed from the Gate House and Newgate, to

Bridewell, to be in case of need punished with torture, doubtless to discover and prosecute Catholic priests.

Tumultuous risings to drive away foreign traders, was by Lord Coke technically termed "Expulit" strangers, and—

April 16, 1593.—Warrant was issued to the Lord Mayor, to *torture* if need be, and he sees fit, to make the person reveal the suspected writer of a lewd and vile ticket set upon a post, purporting determination and intention on the part of the apprentices, to attempt violence on the strangers.

May 11, 1593.—Turbulent conduct of the apprentices of London on the same occasion of discontent.

Divers lewd and continuous libels having been set upon the wall of the Dutch churchyard, to apprehend and examine suspected persons, and to put them to the torture at Bridewell, to be used at such times, and as often as they shall think fit.

November 12, 1595.—Gabriel Colford and his landlord, Thomas Foulkes, tortured with manacles on account of seditious books published abroad.

January 25, 1595.—To examine John Hardie, a Frenchman, for suspicious letters sewn up in his doublet, try him by ordinary torture to get his explanation.

February 1595.—H. Hodges tortured by manacles to find out where £100 was hid in the ground, he having stolen goods and money, and secreted them, the property of Sir H. Bagnall, Knight, attendant about Her Majesty's service.

November 21, 1596.—Eighty Egyptians and wanderers apprehended in Northamptonshire, and tortured by manacles.

Bradshaw and Barton tortured for intention of dealing riotously with enclosures, demolishing churches, instigating several hundred people of lower orders, in Oxfordshire particularly.

The assembly with difficulty suppressed by local magistrates; four ringleaders sent to London; warrants issued. Mr. Bacon and the Recorder of London were to examine the rioters upon such articles as they should think meet, and for the better boulting forth of the truth of their intended plots and purposes; that they should be removed to Bridewell, and put to the manacles and torture.

February 2, 1596.—To the manacles or torture of the rack, Will Thompson, charged with a purpose to burn Her Majesty's ships.

December 1, 1597.—To the manacles. Thomas Travers, stealing a standish of His Majesty, if he would not declare the truth.

December 17, 1597.—Suspicion against the son of an old gentleman, one, Richard Armger, whose body was discovered in the Thames with marks of violence upon it, and a porter of Gray's Inn.

Richard Armger, the son, and Edward Ingram, porter, if they did not confess, to be put to the manacles.

January 4, 1598.—To examine at the Bridewell, and if necessary, to torture by manacles, Richard Denton and Peter Cooper, suspected of dangerous designs against Her Majesty.

A curious commentary on these entries is in another date—14th November.

All the judges being assembled in Serjeant's Inn, Fleet Street, agreed in one, that he (Felton), for assassinating the Duke of Buckingham, might not by the law of the land be tortured by the rack, for no such punishment is known or allowed, by our law.

July 6, 1606.—Derick, the executioner at Newgate, for not branding a culprit, but burning him with a cold iron contrary to order, was punished with twenty-four lashes.

September 22, 1682.—A beadle was appointed to correct prisoners in the house, and those who were to be punished through the streets of the City, instead of the Chapel Beadle, who had performed the office—for the better witnessing the correction, the whipping-post to be raised.

The practice of sending offenders to Bridewell merely to receive corporal correction continued down to a late period. In 1793 as much as £80, 6s. was paid to one of the beadles for flogging prisoners during the previous two years at five shillings each, which gives about 160 as the number of persons annually punished.

Only two instances of corporal punishment occurred, however, during the ten years ending Christmas 1836.

Mr. Martin, in his report on charities, remarks: "It is difficult to imagine how the governors could justify these acts of authority."

Indeed, the powers of police contained in the charter seem to be illegal. Sir Francis Bacon's opinion upon the Charter of Bridewell, was that accusations against people of ill repute are not sufficient without indictment or other matters of record, according to the old law of the land; for Magna Charta maintains that no freeman shall be taken or imprisoned, but by lawful judgment of men of his degree, or by the law of the land.

With regard to the internal economy of the house, it appears that on May 2, 1582, the mill for grinding corn was let to George Green, citizen, and Brown, baker, to take charge of the mill and employ the prisoners on it, for forty years at a yearly rent of £20, to do the work thoroughly, see that the workers did not idle away their time, and to have always corn and grain ready, and should the mill stop for any good reason, 12d. out of the yearly rent per diem to be allowed.

April 30, 1637.—Two beadles were ordered to

daily walk the streets to clear them of beggars and vagrants.

In 1656 it was found that the beadles did not do their work, so the "Corporation of the Poor in London" requested that they might appoint sixteen beadles, and if beggars and vagrants, young or old, were found on apprehension unable to work in Bridewell, to punish them, and to pass them on to their place of birth, or last settlement, and in May 19, 1658, further stringent rules were laid down "that the beadles be most strict in apprehending beggars, and pass on those who ought to be provided for elsewhere; the youngest beadle to punish some by whipping, and others to be kept at work until legally discharged."

"The Treasurer and Governors to examine the women and dispose of them. To have the mill set in order, and to keep at work as many hands as were necessary—presumably so—and if more room was required for a larger number of prisoners, to provide the same."

In February 19, 1673, the marshal's men were reproved for bringing in the lame, blind, and aged persons and young children unfit to be put to labour, and in future were required to punish them and pass them on, and on no account to receive such into the house.

The apprehension of vagrants by the beadles became obsolete in 1785, and an order was made (June 16) that for the future the porter receive no prisoners without a legal commitment by a magistrate.

The scope of the Governors was confined to the interior arrangements and management of the prison. In 1814 the Common Council petitioned the Governors to co-operate with them towards furnishing more accommodation for people of profligate character, but to little purpose; and it was a matter to be regretted, that the prison was not more serviceable to the police. In fact, the state of the prison was anything but satisfactory. The means of classification during the day, and of air and exercise were wholly wanting; beating hemp, picking oakum, grinding corn, drugs, and other kinds of labour had been abandoned, and although some employment was provided for the females, the male prisoners passed almost all their time in idleness.

Bridewell was occasionally used as a State prison. On March 2, 1576, Martin Corbet appeared before the Court, for that the Governors had received a warrant from Her Majesty's Commissioners for Causes Ecclesiastical, and was conveyed to prison until such time as he had satisfied the effect of the said warrant.

January 28, 1656.—James Naylor was committed on a warrant from the Speaker of the House of Commons, dated December 15th previously.

Prisoners were likewise occasionally received on

charges of felony, and detained for safe custody till trial.

January 21, 1642.—It seems that several Turks were then confined in Bridewell waiting their trial at Newgate on a charge of piracy.

November 26, 1658.—Thomas Bullock and another were received on suspicion of stealing goods of the value of £17, to work his jail delivery and be thus sent up. In the reign of Queen Anne, felons convicted at the Old Bailey were sent to Bridewell for punishment; but September 9, 1713, the Governors refused to receive them any more, and memorialised the Secretary of State for that purpose.

CHAPTER VI.

"ELLWOOD'S EXPERIENCES."

THOMAS ELLWOOD, of Amersham, born 1639, and so often imprisoned for attaching himself to the Quakers, in his "History written by himself" of old Bridewell gives the following narrative.

"I was that morning, which was the twenty-sixth day of the eighth month, 1662, at the meeting at the Bull and Mouth, by Aldersgate, when on a sudden a party of soldiers rushed in (of the trained bands of the City) with noise and clamour, being led by one who was called Major Rowsell, an apothecary, if I misremember not, and at that time under the ill name of a Papist. He made a proclamation that all who were not Quakers might depart if they would.

"It so happened that a young man named Dore, from Chimer, near Crowell, in Oxon, came that day in curiosity to see the meeting, and finding me there (whom he knew) came and sat down by me. As soon as he heard the noise of the soldiers he was much startled, and asked me softly if I would not try

E

to get out. I told him no; I was in my place and
was willing to suffer. He turned away and went out.
He that commanded the party gave us first a personal
charge to come out of the room, but we who 'ought
to obey God rather men' stirred not, whereupon he
sent some soldiers to drag or drive us out, which they
did roughly enough. He had gotten thirty-two of us,
and ordered pikes to be opened before us, and the
word to march given, the soldiers making a lane to
keep us from scattering.

"He led us up Martins, and so turned down to
Newgate, where I expected he would have lodged us.
But, to my disappointment, he went on through
Newgate, and turning through the Old Bailey, brought
us into Fleet Street. I was then wholly at a loss to
conjecture whither he would lead us, unless it were
to Whitehall, for I knew nothing then of old Bridewell; but on a sudden he gave a short turn, and
brought us before the gate of that prison, when
knocking, the wicket was opened forthwith, and the
master with his porter ready to receive us.

"One of those who had been picked up in the
street to go with us, happened not to have been
with us in the meeting: this I represented to the
Major, who, incensed at my previous question I
had put to him about a massacre, looked sternly at
me and said, 'Who are you that take so much upon
you? Seeing you are so busy, you shall be the

first man that shall go into Bridewell,' and he thrust
me in by the shoulders.

"As soon as I was in, the porter pointing with his
finger, directed me to a fair pair of stairs on the
further side of a large court, and bid me go up
those stairs, and go on till I could go no further.
Accordingly I went up the stairs, the first flight
whereof brought me to a fair chapel on my left
hand, which I could look into, through the iron
grates, but could not have gone into if I would.

"I went a storey higher, which brought me into
a room which I soon perceived to be a court or
justice room. Observing a door on the farther side,
I opened it, but withdrew from going in, being
frightened at the dismal appearance of the place ;
for, besides the walls being laid all over from top
to bottom in black, there stood in the centre a
great whipping-post, which was all the furniture it
had.

"In one of these rooms judgment was given, and
in the other, it was executed on those ill people, who
were sent to this prison, and then sentenced to be
whipped, which was so contrived that the court might
not only hear, but see, if they pleased, their sentence
executed.

"A sight so unexpected and so unpleasing gave me
little encouragement to rest or even enter, till I
espied on the opposite side another door which I

opened. This led me into one of the fairest rooms
I ever remember to have seen, for it was the dining-
hall of the royal seat or palace of the Kings of
England until Cardinal Wolsey built Whitehall, and
offered it as a peace-offering to King Henry VIII.,
who, until that time, had kept his court in this house
(Bridewell).

"The room, for I lived in it long enough to
measure it, was sixty feet in length, and proportion-
ally broad. On the front side were very large bay
windows in it, and therein stood a large table ; the
floor was covered with rushes for some solemn
festival.

"My thoughts were disturbed by the flocking in
of my other friends—my fellow-prisoners—with whom
I had little acquaintance, having been so short a
time in the city.

"Soon after we had gotten together the master of
the house came up and demanded our names ; this we
need not have done until legally convened before
some civic magistrate, but we, being neither guileful
nor wilful, gave our names simply. It happened that
so great was the storm that fell so heavily upon our
meetings, the prisons were very full of our friends, who
had been apprehended at the several meetings, and
no less care and pains had the authorities to furnish
necessary accommodations and provisions.

"This prison of Bridewell was under the care of two

honest, grave, discreet, and motherly women, whose
names were Anne Merrick (afterwards Vivens), and
Ann Travers, both widows. They provided some hot
victuals, meat and broth, for the weather was cold,
with bread, cheese, and beer, and gave notice to us
that it was provided for all those that had any one to
provide for them.

" For myself, I had tenpence, all the money I had
about me, or any when also at my command. But
' *Natura pauca contenta.*' Hungry as I was, I felt
that I was not included in the invitation, so sat as far
from the table as I could.

" When evening came, the porter told us that we
might have simple eatables, as bread and cheese, eggs
and bacon, &c., from the chandler's shop in the
house, and many gave him money to pay for what
they required.

" He brought me two halfpenny loaves, with which
I regaled myself, reserving one for the following day.
This was to me both dinner and supper, and I had
liked to have gone to bed had there been one ready,
but as there were none of any kind, we walked about
to keep ourselves warm, and sat about all night.
Fortunately some one had bought some candles to
prevent us being in total darkness.

" I made the best of sleeping accommodation by
spreading rushes under the table and using one end
of its frame for a bolster. I, who had no one to look

after me, had to endure this rushy pallet for four
nights, to my intense discomfort; yet, thank God, I
rested well, enjoyed health, and took no cold.

"Many of my companions were released by Sir
Richard Brown, who was a great man at Bridewell,
at the instigation of relations and acquaintances, and,
thanks to the courtesy of one William Macklaw, I
accepted his offer of his hammock whilst I was a
prisoner. Several formed a sort of club and obtained
from Anne Travers agreeable provisions, but my means
were too limited to permit me to join them, for they
judged me by my person, and not by the lightness
of my purse. Yet Providence sent me supply. One
William Pennington called to see me, and desired
me to accept twenty shillings; this I did with thank-
fulness. And he, going to Chalfont to see his brother,
reported my imprisonment, whereupon Mary Pennington by him sent me forty shillings. Soon after I
received twenty shillings from my father, for my support in Bridewell, and the letter he forwarded to Mr.
Wray through my sister I suppressed, the purport
being to get Mr. Wray to intercede with Sir Richard
Brown for my release.

"Acknowledging the goodness of God, I could now
join the club, and I became one of their mess.

"The chief thing I now wanted was employment.
Many being tradesmen, could soon set to work, but I
being a novice, could not be trusted, lest I might

spoil the garments; so I got some from a hosier in Cheapside, and made night-waistcoats of red and yellow flannel for women and children."

It appears that this occupation served merely to pass away Ellwood's time, for he never got a penny for his work other than one crown-piece when he came out of prison, and no more, although he had made many dozens of waistcoats and bought the thread himself.

He relates how one poor fellow, for finishing a pair of shoes on a Sunday, was informed against, and Richard Brown committed him to Bridewell to hard labour at beating hemp. This he refused to do, as he had done no evil, and for the refusal he was cruelly whipped. "The manner of whipping there is to strip the party to the skin from the waist upwards, and having fastened him to the whipping-post, so that he can neither resist nor shun the strokes, to lash the naked body with long but slender twigs of holly, which now bend almost like thongs and lap round the body, and these have little knots upon them, tear the skin and flesh, and give extreme pain."

This poor man was a Friend, and when his tormentors could make no impression upon him, they turned him in with Ellwood and his companions.

With some balsam, his skin, which was dreadfully cut and torn with the rods, back, sides, arms, and breasts were dressed, and after a while got, sound and well.

From the 26th August to the 10th of October they
were kept in prison. They then attended the Sessions at the Old Bailey, but the case was not called;
so they returned to Bridewell till the 29th October.

They all were sent then to Newgate for not taking
the oath of allegiance, which Ellwood cleverly fenced,
for it is to be remembered he was a man of gentle
birth and education, alleging the plea that being
a prisoner, he could not take this oath freely and
without constraint.

The misery endured in Newgate was intense, and
one man died.

Much against his will, an old citizen passing by
was made foreman of the inquest, and desiring to
see the place where the dead man had been kept,
was astonished at the miserable, pestilential place.

The next day Sir William Turner, one of the
Sheriffs, caused all those who had come from Bridewell to return thither, and this they did *without a
keeper*, their word being alone taken.

"When we were come to Bridewell, we were not
put up into the great room in which we had been
before, but into a low room in another fair court, which
had a pump in the middle of it; and here we were not
shut up as before, but had the liberty of the court to
walk in, and of the pump to wash and drink at.

"We could have gone away, as there was a passage
out into the street, but we were true and steady

prisoners, and looked upon this liberty as a kind of parole upon us."

They were treated with considerable leniency and indulgence until the court sat at the Old Bailey again, when they were all called to the bar, and without further question discharged.

CHAPTER VII.

REMINISCENCES OF THE PRISON.

NED WARD, in his " London Spy " (A.D. 1703), gives
an account of Bridewell, but it is not very enter-
taining.

His experiences take the form of unsavoury dia-
logues between the prisoners and himself or his
companions. Yet one poor fellow was detained
for being unable to pay his fees, which amounted
to five groats. Ward remarks, " Bless me ! thought
I, what a rigorous uncharitable thing is this that so
noble a gift, intended, when first given " (alluding to
the Hospital), "to so good an end, should thus be
preserved, and what was designed to prevent people
falling into misery through laziness or ill courses,
should now be corrupted by such unchristian confinement, or to starve poor wretches because he
wants to satisfy the demand of a mercenary Cer-
berus when discharged by order of the court : such
severe, nay, barbarous usage is a shame to our laws,
an unhappiness to our nation, and a scandal to
Christianity."

He praises the magnificent, noble buildings which composed the prison or penitentiary, and, in going to the female side, is struck with its occupants. Some seemed so very young, to be brought thus early into a state of misery, and others so old that one would think the dread of the grave and thoughts of futurity were sufficient to reclaim them from vice.

Sick, amazed, and tired with the behaviour of these unhappy culprits, who had neither sense of grace, knowledge of virtue, fear of shame, or dread of misery, Ward and his friend proceed to the court-room, where they witness the trial and punishment of a young woman, who had to strip to her waist and be flogged until the master of the tribunal let his hammer fall to show that sufficient punishment had been administered; giving rise to the old expression, so often used, when this same whipping took place, by the prisoners, " Knock, good Sir Roger, knock ! "   The moral which the author of the " London Spy " draws from the before-mentioned scenes is as follows :—" According to my real sentiments, I only conceive it makes many bad women, but that it can in no measure redeem them; and these are my reasons,—First, if a girl of thirteen or fourteen years of age, as I have seen some others, either through ignorance or childishness of their youth, or unhappiness of a stubborn temper, should be guilty of negligence in their business, or prove headstrong, humoursome, or obstinate, and through an ungovernable temper take pleasure to do things in disobedience to the will of their master and mistress, or be guilty of a trifling wrong or injury through inadvertency, they have power at home to give them reasonable correction without exposing them to this shame and scandal, which is never to be washed off by the most reformed life imaginable, which unhappy stain makes them always shunned by virtuous and good people ; also will neither entertain a servant nor admit of a companion, under this disparagement, the one being fearful of their goods and the other of their reputation, till the poor wretch by her necessity is at last drove into the hands of ill persons, and forced to betake herself to bad conversation, till she is insensibly corrupted and made fit for all wickedness.

"Secondly, I think it is a shameful indecency for a woman to expose her naked body to the sight of men and boys, as if it was designed for other purposes than to correct vice or reform manners ; therefore I think it both more modest and more reasonable they should receive their punishment in the view of women only, and by the hand of their own sex.

"Thirdly, as their bodies by nature are more tender and their constitutions allowed more weak, we ought to show them more mercy, and not punish

them with such dog-like usage, unless their crimes
were capital."

The following lines are added :—

> " 'Twas once the palace of a prince,
>   If we may books confide in,
> But given o'er by him long since
>   For vagrants to reside in.
>
> The crumbs that from his table fell
>   Once made the poor the fatter,
> But those that in its confines dwell
>   Now feed on bread and water.
>
> No venison now, whereon to dine,
>   No fricassées, no hashes ;
> No ball, no merriments, or wine,
>   But woeful tears and slashes.
>
> Where once the king and nobles sat,
>   In all their pomp and splendour,
> Grave City grandeur nods its pate,
>   And threatens each offender.
>
> Unhappy their ignoble doom,
>   Where greatness once resorted ;
> Now hemp and labour fills each room,
>   Where lords and ladies sported."

Only a few more words are needed respecting
Bridewell as a prison.

"Let the sorrowful sighing of the prisoners come
before thee," are the words which John Howard the
philanthropist introduces, in his work upon "Lazarettos
Abroad and Prisons in England," 1789.   In referring

to Bridewell, he says :—"No alteration but the venti-
lators taken down.   Each sex has a workroom and a
night-room.   They lie in boxes with a little straw on
the floor.   The prison not being strong, the men were
in irons, some picking oakum, and others were making
ropes, which is a new and proper employment.   Mr.
Hardwick, a hemp-dresser, has their labour, and a
salary of twenty guineas a year.   Allowance, one
penny loaf each, and four days in the week ten
ounces of beef without bone, &c.   The allowance
for persons constantly employed, is not too much,
but would it not be better if they had less meat
and more bread?   The prison wants white-washing,
and the men's night-room more light and air.   At
my first visit two men were in the infirmary ; at my
last, only one.

    1787, November 6,   Men, 26 ; women, 25.
    1788, September 13,   .   Men, 19 ; women, 10.

" There are many excellent regulations in this establishment.   The prisoners have a liberal allowance,
suitable employment, and some proper instruction ;
but the visitor laments that they are not more
separated."   He adds in commendation, "In winter
they have some firing, the night-rooms are supplied
with straw ; no other prison in London has any
straw or bedding."

There are, very properly, solitary cells for the

Bridewell boys, in which one was confined and employed in beating hemp.

Vagrants and others committed to the prison in

Year.		Prisoners.
1783 .	. .	1597
1784		2956
1785 .		612
1786	.	716

These numbers are from accounts made up every Easter.

HEPWORTH DIXON, writing more than a century later in his "London Prisons," 1850, thus refers to Bridewell:—"At present it is a sort of House of Correction to the City of London. Summary convictions and apprentices sentenced to solitary confinement are sent hither, but not many of the latter are to be found there. Every care is taken to prevent communication with vagrants or others also occupying the building. Troublesome as some of the lads are, they are not to be confounded with the felons; they probably proved good stuff after all.

"As a House of Correction it was bad, unhealthy, the apartments small and straggling, ill arranged, and no sort of superior supervision worthy the name; cells and corridors dark and confined, insufficient light and air; yet it is superior to Giltspur Street Compter, and Horsemonger Gaol, and, on account of separate sleeping apartments, better than any gaols in

London except Pentonville and the Middlesex House of Detention.

"The numbers were—men, 70; females, 30; chiefly under sentences of three months' hard labour, which consisted of the tread-wheel and oakum-picking, most kept at the wheel, and the straining figure of the criminal may be dimly seen. The only sound apparently is the dull soughing of the wheel; and the dark shadows toiling and treading in a journey which knows no progress, force on the mind, involuntary sensations of horror and disgust.

"The system of discipline pursued is a mere mockery of the silent system. Communication is forbidden during hours of work, but not prevented. The walkers of the wheel are commanded not to talk, but from the straggling nature of the building and the paucity of prison officers, complete inspection and control are out of the question, and practically they talk just as much as they think proper: as when at work, only a thin partition separates one from another. Nothing less than the presence of a warder could prevent them. They who are not sentenced to hard labour are confined in the opposite wing of the prison. Underground are two small miserable cells, the day-rooms of this department. They are very cold and damp, consequently fires have to be kept in them, a circumstance fatal to all discipline.

"In each of these rooms at the time of our visit

there were eight or ten prisoners shut up, picking oakum. They were quite alone, that is, no officer was with them in the room. Occasionally they receive a visit, but by far the greater portion of the day they are quite alone, talking over the fires, and instructing each other in their favourite arts. How different this, to the workrooms of Millbank or Coldbath-fields !

" There is no school in this prison. This is the fitting climax of its many faults, the crowning absurdity of the whole system of mismanagement. If book-teaching be absolutely required anywhere, if it promise to be successful to operate for good anywhere, surely it is here in Bridewell. Being summary convictions, the inference is that persons come here, to get their initiation into the prison world, and the fact is so generally.

" The ill-directed youth of the city, who commits his first petty offence, is most likely to be sent hither. Upon the impressions which he takes away, may depend the entire future of his existence for good or evil; in a course of reform or a career of guilt, his incarceration in Bridewell is the starting-point. One thinks with pain and sorrow of the education which such a youth must get here now, and of the direction most likely to be given to his energies by the persons he will meet with. Three months' imprisonment here is enough to ruin any child for life. The boy must have powerful elements of good in him who can leave

F

it no worse for ninety days' contact with its contamina-
tions.   Instead of subjecting the unfledged criminal
to the pollution of his unrestrained intercourse with
offenders worse than himself, every care should be
taken here, upon the threshold of his career, to lead
him back from the fatal path into more respectable
and honest courses.   The negative act is not enough.
He should not alone be kept from peril; he should
also be put into the way of good.

"Two means are patent for this purpose—work and
teaching.   The work should be severe but useful,
such as a man in whom it was sought to foster habits
of self-respect, might be asked to do.   The instruction
should be sound and regular; what the criminal mind
wants most is discipline.   Formerly there was a school
at Bridewell; it has for unknown reasons been given
up.   Fatal mistake!   If anything could atone for the
faults of the City Bridewell, it would be the institution
attached to it, called the House of Occupation, in
St. George's Fields.   This is, in fact, an industrial
school; and has about 200 inmates, half male, half
female.   It is not a criminal establishment.   The
majority of its scholars have not been in prison; the
minority have, in Bridewell.

"Children who are idle merely, disposed to be
troublesome to their parents and to the community,
are taken in, educated and instructed in a trade, and
after several years of careful training are placed in

situations, or permitted to go home to their parents on the latter making proper application.

"The instruction given to them is sound and practical, the discipline enforced strict, but not rigid, and the general result highly successful. The boys are taught trades. At present there is one or more, learning each of these useful employments—engineering, painting, tailoring, shoemaking, masonry, binding, baking, carpentry, rug-making, rope-making. The girls are being taught every species of domestic work —washing, sewing, cooking, ironing, knitting, &c.

"Great care is also taken with the education of their minds; they are said to make admirable domestic servants, and very rarely indeed does one turn out ill. They are in great request, there being usually from twelve to twenty applications for servants on the books of the institution.

"As they are ready they are put out from Bridewell; the Magistrates have a power of removal to this House of Occupation, being one of the first Reformatory Schools established, and by the change of scene, this removal from old haunts, old comrades, and old occupations, hundreds of poor boys are placed in a position for becoming useful and productive, instead of dangerous and expensive, members of society."

CHAPTER VIII.

REPORTS ON PRISON, 1855-1887.

An interesting communication was made in 1886 (5th November) by General Adams, late Governor of the prison, who wrote :—

" I was appointed to the office of Superintendent in November 1847, and held it till March 1855, when the prison was closed except for the reception of refractory City apprentices committed by the Chamberlain.

" On my services being no longer required, the Governors granted me a very liberal pension, which I still enjoy. The staff of the prison consisted, besides myself, of the chaplain, one chief, and four assistant warders on the male side, and of a matron and two female warders for the female prisoners.

" The prisoners were committed by the Lord Mayor and Aldermen, and the period of their confinement was from three days to three months.

" They were principally pickpockets, and very generally gave as their place of abode the New Cut or Mint Street, in the Borough. Most of the others were vagrants sent to prison for begging in the streets.

"Many of the prisoners were well known to the officers, having been frequently inmates of the 'Old House,' as they called Bridewell.

"Hard labour was carried out by the treadmill, which, communicating with the adjacent mill, was the motive power for grinding the corn for the use of the two Hospitals and House of Occupation.

"Such prisoners as were unfit to undergo hard labour were employed in chopping wood, or any light work about the prison.

"The females did all the lavatory work, and a few of them assisted in preparing the food.

"The conduct of the male prisoners was generally good, and they were always very respectful.

"The females were not unfrequently refractory; so much so as to oblige the matron to call for the assistance of the warders on the other side. On these occasions the women would scream until they were quite hoarse, and tear up their clothes and bedding into shreds, and smash the windows of their cells."

The Governors' report in February 1855 was as follows:—

"The Governors having, at an early period of the year, decided upon the expediency of closing the prison of Bridewell and of applying the large revenues of the Hospital to more useful and beneficial purposes, solicited the concurrence of the Court of Aldermen in that arrangement; and as they practi-

cally acquiesced in the proposal by ceasing to commit
any prisoners subsequent to the month of April, the
prison may almost be said to have been closed from
that time, although at a later date the City apprentices
committed by the Chamberlain have been received.

"The above statement was necessary, in order to
explain the cause of the very small number committed
in 1854 when compared with those in former years;
but it may likewise be observed, that although the
committals of the past year extend over a period only
one-third of that in previous years, the relative proportion is by no means the same, but was much less than
in any year that preceded it.

"In the years 1851, 1852, and 1853, the number of
committals from the 1st of January to the 30th April,
was respectively 394, 278, and 418, whereas in 1854
it was only 65.

"Unfortunately we are not warranted in attributing
this decrease to any diminution of crime in the Metropolis, but rather to the preference evinced by the
City Magistrates to commit prisoners to the House
of Correction at Holloway, in which it is supposed
that the modern system of prison discipline can be
carried out with greater efficiency, than in the old and
obsolete building of Bridewell.

"The number of prisoners committed during the
past year amounted, as above stated, to 65 prior to
the 30th April, and the subsequent committal of two
apprentices gives a total of 67.

"Of these, five were City apprentices committed by the Chamberlain.   The remainder were chiefly either pickpockets or misdemeanants, of which last class of offenders the proportion was somewhat greater than in former years ; but their detention in prison was generally only until their friends could procure the means of paying the fines, in default of which they were committed.

"The number of juvenile offenders was twenty-four, being about one-third of the whole ; and of these five were recommittals.   Their ages varied from eight to seventeen years.

"The health of the prisoners was as usual very good, and their conduct whilst in prison satisfactory.

(Signed)   " E. ADAMS, Capt.,
                                        Superintendent.

" BRIDEWELL HOSPITAL,
    5th February 1855."

There are many of the old commitment warrants still in existence, dating from 1828 to 1853.   One or two have been selected from among the waste paper at Bridewell, as illustrating the offence and punishment, the chief offences being those of vagrancy, idleness in the apprentices, indecency, and thieving, and the term of imprisonment ranging from seven days to three months.

London to wit.

To all and every the Constables and other Officers of the Peace for the City of London and the Liberties thereof, whom these may concern, and to the Porter of Bridewell Hospital, London.

CITY ARMS.

THESE are in his Majesty's name to command you and every of you the said Constables, forthwith safely to convey and deliver into the Custody of the said Porter the Body of THOMAS WRIGHT, being charged and convicted before me, one of his Majesty's Justices of the Peace in and for the said City and Liberties, by the Oath of THOMAS DENNET, *with wandering abroad in the open air, not being able to give a satisfactory account of himself, on Thursday the fourth instant, in a certain place called Mansion House in this City,* which being proved before me, I do convict *him* and do adjudge *him* to be a Vagabond within the intent and meaning of the Statute made in the fifth Year of his Majesty King *George* the Fourth, intituled *An Act for the punishment of idle and disorderly Persons, and Rogues and Vagabonds, in that part of* Great Britain *called* England, whom you the said Porter are hereby required to receive and *him* to keep and set to hard labour for the space of *One Month* until *he* shall be discharged by due course of Law; and for your so doing this shall be to you and each of you a sufficient Warrant. Given under my Hand and Seal, this *sixth* day of *December* 1828.

WM. THOMPSON, *Mayor.*

To all and every the Constables, and other Officers of the Peace for the City of London, and the Liberties thereof, whom these may concern, and to the Porter of Bridewell Hospital, London.

London to wit.} THESE are in His Majesty's name to command you and every of you forthwith safely to convey and deliver into the custody of the said Keeper the Body of JOHN GOSLING, *he* being charged before me, one of His Majesty's Justices of the Peace in and for the said City and Liberties, by the Oath of *Henry Kitchennan, for that the said John Gosling, being an apprentice to the said Henry, hath committed divers misdemeanors against his said master, and in particular for that the said John hath run away and absented himself from the service of his said master for the space of five days contrary to the statute, &c., of which I have convicted him, and for his said offence have adjudged him to be imprisoned for one month,* whom you the said Keeper are hereby required to receive and *him* in your Custody safely keep *to hard labour for the space of one month until he* shall be discharged by due course of law: and for your so doing this shall be to you and each of you a sufficient warrant. Given under my hand and seal, this *Twentieth* day of *January* 1834.

C. FAREBROTHER, *Mayor.*

The practice of committing to Bridewell for safe custody had long been discontinued in 1837, and no tried prisoners had been received since 1828.

Those in the prison in 1837 were :—

1. City apprentices committed by the Chamberlain for misconduct.
2. Ordinary prisoners summarily convicted by the Lord Mayor and Aldermen.

The warrants by the Chamberlain were and are even now directed to the porter or beadle, according to ancient practice. The confinement is solitary; there are six cells, and the beadle has charge of them.

The Vicar of St. Bride's has notice of the reception, and visits the apprentices daily whilst in the cells.

A curious entry occurs in the Vestry Book of St. Bride's parish :—

" *At a Vestry held February* 26, 1661.

" Upon the petition of younge Jenninges wife, her husband abusing her, the churchwarden is to chide him for it and advise him to behave otherwise; if he doth not amend, then a warrant to goe out against him that he may be committed to Bridewell till he gives security for his good behaviour."

There were 52 recalcitrant apprentices in the House in 1836; in 1886 and 1887, only 17.

The punishment is not considered a criminal conviction, nor does the Chamberlain commit unless the offence is really discreditable, as neglect of work, absconding, playing truant, &c.

CHAPTER IX.

ART-MASTERS AND APPRENTICES.

IT seems that the manufactures originally introduced into the House were to be carried on, on account of the Hospital, and it is probable that the more complicated operations were performed by persons receiving wages for their labour, and hired into the House to work up the materials prepared by the inmates, the due execution of the work, being secured by the selection for each trade, of governors particularly acquainted with the requisite details.

In the ordinances of 1557 the word "apprentices" occurs. And in a report of a committee in 1657 it is stated that by several leases which were executed about that year (1657), houses and apartments within the Hospital were demised to several persons who were appointed "taskmasters and taskmistresses" for the management and improvement of different manufactures, and bringing up "apprentices to the same."

There is extant a copy of an indenture dated 22d November 1577 (19th of Elizabeth), whereby certain houses were demised by the Mayor, Commonalty, &c., to Richard Matthew, at Bridewell, for ninety-two years, with a covenant that he and his executors, &c., should instruct and bring up youths in the trade of making knives, steel buttons, blades, &c., and the lease to be void if the premises should come to any person not free of the city.

Another instance occurs of such a lease, for in 1585 the Commonalty demised certain houses to Thomas Dowleyn, cutler, who covenanted to receive, take, instruct, and keep at work, as his daily servants, upon some decent and good handicraft, all such poor vagrants as should be sent by the Governors, and that he should lodge and board them, and not permit them to escape.

The precise date of the introduction of art-masters and legally bound apprentices cannot be ascertained.

An old weekly account book, says Mr. Bowen, a former chaplain of Bridewell, contains an entry, 23d May 1594, of a payment of 12s. 8d. for six pairs of indentures of apprenticeship for boys bound with the glover.

In February 1597, again, that one Exton and family, be allowed 26s. 8d. per week for teaching boys pin-making.

At a court held February 8, it was ordered that the Treasurer should examine the accounts of the late "art-masters of the pinners," and this appears to be the first time the word occurs in the records.

The first instance of regular binding occurred at a
court held 28th March 1598, when Nicholas Ling,
churchwarden of St. Clement's, Eastcheap, paid £5
"to place Thomas Scarlet, apprentice with Thomas
Ellis, the glover, for seven years."

Ellis was paid 12s. more, and 30s. for diet since
"Hallentide" last, and the court discharged the ward
of Candlewick, of the charge of Scarlet during his
term of apprenticeship.

By an order, October 10, 1599, the Governors
agreed to receive parish children and children of
freemen, to be taught some trade; and Richard
Brooke, a fustian weaver, was allowed a house rent
free and to keep ten apprentices.

March 18, 1600, it was ordered, with full consent
of the Lord Mayor, that Aldermen might send to the
Hospital, parish children within their wards, to be
placed under some artificer.

December 8, 1606, Churchwardens of the parishes
of St. Sepulchre's, St. Giles's Cripplegate, St. Bride,
St. Botolph, Aldersgate and Aldgate, to be allowed to
send any poor boy, to be set apprentice.

This system of apprenticeship was fostered and
encouraged, by the bequests of Locke, Fowke, and
Palmer.

In March 1644, it was recommended that vagrants
found in the streets, should be brought to Bridewell;
and that the small children born in the City, and

not able to move, be kept at Christ's Hospital, to be reared and taught, and on attaining the age of twelve years be sent back to Bridewell to be employed in some good occupation.

About the year 1671 the better education of the apprentices was considered, and a school established in the house for them.

In 1720 the art-masters and apprentices appear a very numerous, and also a very disorderly part of the Hospital.

The insubordination and irregularities that had arisen were occasioned by the free liberty they possessed to quit the precincts of the house, and the custom existing from an early period, of permitting the engine of the Hospital to attend all fires.

The Bridewell engine was noted for its efficiency and for the courage and dexterity of the apprentices, called "Bridewell boys;" but this practice resulted in frequent injuries, drunkenness, and debauchery.

Hone mentions in the "Everyday Book:"—"On the 13th November 1755, at a Court of Governors of Bridewell Hospital, a memorable report was made for the Committee, who inquired into the behaviour of the Bridewell boys at Bartholomew and South-wark fairs, when some of them were seriously corrected and continued, and others, after punishment, were ordered to be stripped of the Hospital clothing and discharged. The Bridewell boys were, within recollection, a body of youths distinguished by a particular dress, and by turbulence of manners. They infested the streets to the terror of the peaceable; and, being allowed the privilege of going to fires, did more mischief by their audacity and perverseness than they did good by working the engines." It is only right to add that some improvement must have taken place in their manners; for Hone concludes by admitting, that "the Bridewell boys at this time" (the book was published in 1827) "are never heard of, in any commotions, and may be regarded therefore as peaceable and industrious lads." Nevertheless, their attendance at fires seems to have been dispensed with.

In 1792 a strong opinion prevailed that the system of art-masters and apprentices was extravagant and useless, and on the 14th June, it was resolved that it should be abolished, and notice to quit was actually given to the art-masters. This was modified by a subsequent report.

In 1798 there were no apprentices in the House.

In 1799 a lengthy report contained suggestions for the better classification of the inmates of the House, *e.g.*, for the reception of destitute persons discharged from prisons and hospitals, and for the institution of a distinct establishment as a "School of Occupation" for the uneducated children of the Metropolis, and urging the entire inutility of the art-masters and apprentices.

The report was not confirmed.

Admission of apprentices was resumed after an inquiry in May 1799, and the report confirmed, and they were chiefly taken from the boys who had received at Christ's Hospital the lower grade of education.

None were appointed from Newgate, or taken from the streets, yet nothing could be more pernicious than assembling together, a large number of young men within the same building with abandoned and dissolute prisoners of both sexes; for though they might be completely separated, the inmates were lowered in public estimation, and their prospects of future employment prejudiced.

CHAPTER X.

REPORT OF A.D. 1818.

THE Report of the Committee of the House of Commons, published in 1818, condemned the existing state of things.

" Bridewell is that Hospital of the three named in a Charter of King Edward VI., nominally devoted to the employment of the idle and disorderly, but in practice effecting neither. Although ostensibly a House of Correction, no attempt is made to reclaim the prisoners or to correct them, except by administering corporal punishment, which is left in a great measure to the discretion of the porter. No employment of any description is provided. A few women work at spinning-machines, but the men for the most part saunter about.

" The revenues, amounting to £7000 a year, are not well applied. However defective former arrangements may have been, the present are useless. The evil has grown by gradual progress. Vagrants are, it is true, received and fed for a few days until they can be passed on to their proper settlements. The disorderly are confined, but ridicule the correction.

"Apprentices are admitted and taught by art-masters, but they might better gain a knowledge of their trades elsewhere at far less expense, and a minute inquiry is recommended."

The Court held November 6, 1818, appointed a Committee, who, in their report of the 20th October 1819, confirmed the report of the Parliamentary Committee, that the prison, both in structure and management, was defective, and that a radical change was necessary; but that doing away, or immediate interference with, the art-masters was to be deprecated.

It appeared that in March 1819 there were twenty-eight apprentices in the house, out of ninety-three that had been bound during the twenty-one years previous to that date. On June 22, 1821, a resolution was adopted—"That it is the opinion of this Court, that the prison of Bridewell be altered, to admit of proper classification and superintendence, and that it will also be proper, so far as the revenue and the altered state of society will admit thereof, to restore the Hospital to its original condition of a House of Occupation."

Males and females to be received as follows :—

(1.) To be instructed in useful handicraft, trade, or occupation, whereby an honest livelihood may be obtained for the future.

(2.) The unemployed, whether committed by the magistrates as idle, disorderly, &c., or being apprentices *committed by the Chamberlain*, or received at their own desire, to be usefully employed until discharged.

(3.) Prisoners, quit at the Sessions, desiring temporary refuge and maintenance, to be employed usefully during their continuance in the Hospital.

Proper rules and regulations were to be drawn up for the classification and employment of those committed, and those admitted at their own request; and that a rather lower scale of wages than is usual, be given to the latter for work done.

That the art-masters be termed taskmasters, and be freemen of the City.   That a school be established for the general instruction of the children in religious and moral duties.

A superintendent and other officers, as keeper, matron, and turnkeys, were appointed, and considerable alterations were made, treadmills, &c., being provided.

In 1835 a new wing was added to the prison.[1]

The House of Occupation was not resolved upon until March 7, 1828.

The resolution was as follows :—

---

[1] Benge, the beadle, December 21, 1887, told the author that a very decent man conversed with him not long since, who had been an apprentice in Bridewell, and that it was the happiest time of his life.

A mother who had been a prisoner long ago, told Benge, that when the prisoners were discharged, a loaf of bread was given to each, and this they always stuck on the railings to show their contempt.

"That a new House of Occupation for the reception of destitute objects of both sexes, should be provided. That destitute people committed to the House of Correction at Bridewell, and disposed to work, should be received at the expiration of their time of confinement, and remain at the discretion of the Governors."

A lease was granted for sixty-one years, at a yearly rent of £200, from August 1, 1828, for the House of Occupation, on three acres of land in St. George's Fields belonging to Bethlehem Hospital, and a building was erected at a cost of £14,900, and opened October 1830.

It only remains to follow the history of the *House of Occupation* till such time as it was altered and called "King Edward's Schools."

The Report of 1836 says, that youth only, of both sexes, are admitted, and the inmates divided into the following :—

(1.) Persons merely destitute.

(2.) Persons guilty of misconduct. And this class included those, not convicted of offences against the law, but of idle, dissolute, vicious, and bad habits, and uncontrollable by their parents ; young women who had gone wrong, and were desirous of returning to a better life ; young persons of both sexes desirous of amendment, after committal on summary convictions, and those convicted at Sessions on whom judgment had been respited.

The age was from eight to nineteen, and preference given, *cæteris paribus,* to those discharged from
Bridewell.

Great attention was paid to their cleanliness, diet,
and clothing; the trades taught (besides the performance of all the necessary economies of the institution),
appear to be brewing, baking (the flour came from
the mill at Bridewell), ropemaking, bootmaking, and
tailoring; and one notices that education, particularly
religious instruction, of which the inmates were very
ignorant, had prominent attention.

The total admission of both sexes from 1830 to
1853 were 1632, 842 of whom were males and 790
females, and of these 305 had been in prison, or had
been sent from prison. 185 males and 120 females.
The Rev. Mr. Garrett, chaplain, writes as follows in
his report, dated March 15, 1854 :—" It is a very
remarkable and most interesting fact, and I trust,
gentlemen, you will pardon my again drawing your
attention to it, that the number of those young persons of either sex who have left this House under
unfavourable circumstances, and have taken a position
in society honourable to themselves, thus reflecting no
trifling degree of credit upon the labours of those, to
whom you have confided so responsible a charge, very
greatly preponderates over those, who have again fallen
into a vicious course of life, and this has been more or
less the certain results of our inquiries from year to year.

" This fact evidently arises, not so much from a
separation from their former associates (for each com-
mittee-day adds to the amount of moral delinquency
already within the walls), but is rather to be ascribed
to the occupation of that time, which was once a
weapon of mischief in their hands, in the task of
storing their minds with useful knowledge, of instructing them in their duty to God and man, and the
providing them with industrial pursuits, which, when
they step again into the world, will afford them the
means, with God's blessing, of obtaining an honest
livelihood, and also a way to escape, when trials and
temptations assail them."

Mr. Garrett retired from the chaplaincy of Bethlehem and the House of Occupation, in January
1856, after twenty-two years' service. When the
Rev. Edward Rudge made his first report in the
following January, he gives the number of inmates
as 212,—114 boys and 98 girls, but the usual average "rarely falls short of 220 ;" and further remarks
" that much of the evil sought to be cured by means
of the House of Occupation, might have been
avoided altogether by a more conscientious exercise
of parental responsibility."

" Nor does the poverty, or even the ignorance, of
the parents, at all times render this neglect in some
measure excusable. Those who are best acquainted
with the humbler classes, will bear ready testimony

to the struggles made by the industrious and well-principled among them, to appear in decent apparel themselves, and to get their children properly clothed and educated.   The neglect is often the greatest amongst the well-paid and intelligent, who spend most upon personal and selfish indulgences, and least upon home comforts and the education of their children.   It is thus that children become undisciplined, and the parents, willingly or otherwise fostering them in evil courses, are by such means relieved of the charge of them altogether.   The reformatory movement is a hopeful feature of the present day, but the danger just presented is apparent, and suggests caution."

"When I first knew the House of Occupation," he adds, "it was partly a Reformatory School, and partly what I will call a Preventative School, *i.e.*, a school for such destitute children as were in danger, from their unprotected state, of falling into crime. No classification was attempted, nor indeed was possible, and the effect of associating unconvicted children with others who had been convicted, and that more than once (one boy of the age of fourteen, admitted from the City Prison at Holloway, had been convicted of pocket-picking seven times) must have been most injurious to the former.   The Governors had no power to help themselves; by the terms of their charter they were bound to admit both classes.

The criminal children were most difficult to manage; they were constantly trying to escape; skilful in the art of making skeleton keys, and adepts in prison slang.

" Of course they never saw the outside world from the day of their admission to the day of their discharge. They were treated as kindly as circumstances permitted; but I must say that I look back to my first experience of the House of Occupation with anything but pleasurable feelings."

Again, in 1858, the chaplain writes : — "Your House of Occupation differs from the modern reformatory school in these two particulars,—first, that conviction of crime before a magistrate is not a necessary qualification for admission ; and, secondly, that the inmates are neither received nor detained, against their own wishes. But facts show that our real and actual work is very similar, and the Governors may fairly claim the credit of having for many years acted as pioneers to that philanthropic movement for the recovery of the juvenile delinquent, to which a vigorous impulse has in these days been given." A great many boys at this period were received into the Royal Navy. Of 140 boys and 68 girls discharged, 62 of the former entered the Royal Navy, and 12 the merchant service ; and of the latter, 50 went into domestic situations ; and the tone of the girls was much improved under the firm but kind discipline exercised by a new matron.

CHAPTER XI.

KING EDWARD'S SCHOOLS.

IN 1860 the new scheme for the regulation of the charity came into operation. Alterations in the qualifications for admission were made. The name was changed from that of the " House of Occupation " to the more appropriate one of

" KING EDWARD'S SCHOOLS."

The age of admission, hitherto from 13 to 16, was lowered to from 12 to 15, and even in some cases to from 10 to 12. Nor were the admissions confined to the residents in the City of London, the county of Middlesex, and the borough of Southwark, thus rendering the institution not a local, but in every sense a national one. Its character as a school for the prevention of crime rather than for the reformation of juvenile criminals, was more clearly defined, and no criminal children will be received except under peculiar circumstances; and if any be admitted of this class, the proportion not to exceed one-sixth of the

whole number of inmates. Mr. Rudge greets these
features of the new scheme as most excellent. For
the association of children steeped in crime and used
to prison ways, with those who are simply destitute
and unprotected, is manifestly unfair, and calculated
to exercise a baneful influence upon the future course
of life of the inmates generally.

The reproach of having belonged to the school will
exist no longer. "When I became your chaplain,"
continues Mr. Rudge, "five years ago, the majority of
the inmates were criminals. Out of those admitted
last year (1860), which exceeded 200, only 16 boys
and 4 girls had been convicted of crime : 108 boys
were discharged to the Royal Navy and 44 girls to
situations." An officer commanding one of H.M.'s
ships writes :—"It gives me very great pleasure to in-
form you that the lads from King Edward's School who
have joined the Navy in this ship have always behaved
much to my satisfaction. I only regret that, in consequence of the number of applications we receive from
other charitable institutions, we cannot afford to take
more from you." 166 boys and 36 girls, who formerly
belonged to these schools, attended the Commissioners
with certificates of good character and received their
£1 reward, in accordance with the excellent rule of
the institution : 86 boys and 19 girls for the first
time, 48 boys and 10 girls for the second, and 30 boys
and 7 girls for the third and last time. On the 3d

March 1865 the first stone of the new school was laid at Witley, near Godalming, Surrey, by the President, Alderman Copeland (the silver implements used in the customary formalities were presented some years later to the Governors by one of the late Alderman's sons), and the separation of the boys from the girls completed early in 1867; the new buildings being formally opened on the 5th April.

At the end of the year 1868 there were 107 boys and 110 girls in the schools. The intention of the new school at Witley, was for the accommodation of 150 boys, and an equal number of girls at the old school at Southwark, and these numbers were gradually attained. In 1876 there was a total of 311 children in the schools. The Admiralty having raised the standard, and parents proving so unwilling to allow their boys to go to sea, Mr. Rudge bitterly laments his inability to get lads into the Royal Navy. He records that a former inmate, who had left eight years previously for the Royal Navy, had been in the Arctic expedition, and gave on a visit to his old school a most interesting account of his adventures.

Of the eighteen saved when the *Captain* capsized in the Bay of Biscay, three of them were old King Edward School-boys. The author saw the *Volage* in the Portsmouth Roads when, with Mr. Rudge, endeavouring to get some lads taken on board the

St. Vincent training-ship. It was the *Volage* which brought home those who were saved, and the loss of the *Captain* had a bad effect at the time, these huge ironclads being termed by the tars "iron coffins."

Meanwhile, the girls' school had been progressing admirably. The number of applications for girls for private families was in 1877 greatly in excess of the matron's means of supply, and it is most satisfactory to record that in nearly every instance these applications are the result of the observed efficiency of the King Edward School-girls—one lady recommending the school to others having vacancies in their households.

Those of the Governors who have gone over the establishment (which is at all times open to their inspection), and who have observed the modest and cleanly appearance of the inmates, and the careful manner in which they are trained in all the various branches of domestic work, will feel no surprise at this result. Further, in all the little trials and difficulties which they have to encounter at the first start in life, the girls are encouraged to apply to the matron for advice and direction, and correspondence with them and visits to them in their places, whenever a personal interview seems desirable, occupies a considerable portion of the matron's time—time which, however, is well employed; for I attribute to this constant supervision, this kindly interest

in their welfare, even after they have ceased to be
scholars, no small portion of that success which
this branch of the institution has undoubtedly obtained. It must be remembered that the fact of
keeping touch with those discharged for three years
after their leaving the school is a most important
element in their general behaviour. The system of
rewarding for good conduct in situations, whatever
they be, so long as the chaplain and matron is cog-
nisant of their being respectable, by £1 each year,
has been proved most efficacious.

About the management and internal life of the
schools it is hardly necessary to say much. Two
most excellent features are noticeable, the praise for
which is due to the Governors, and not to the execu-
tive. The first is the absolute impartiality with which
the candidates for admission are selected by the sub-
committee of Governors—the most destitute cases
being invariably preferred, and thus the objectionable
system of canvassing, with its expense and frequent
disappointment, entirely avoided. The second is the
admirable rule of giving to former inmates who
attend the committees with good characters from
their employers rewards of £1 for three years. Not
only does the pecuniary benefit, act as an incentive to
good behaviour, and as a barrier against a restless
change of place, but it gives an opportunity of
renewing acquaintance with old scholars, and afford-

ing such advice as circumstances might seem to require.

The numbers in the schools gradually increased, and in 1881 a total of 436 is reported—216 boys and 220 girls—and the enlargement of the premises, both in London and Witley, was completed, and the numbers permitted as a maximum was 240 of either sex.

The Lord Mayor in 1880 (Sir F. W. Truscott) visited the boys' school on the annual examination-day; his example was followed by Sir Henry Knight in his year of office in 1883, and by Sir Reginald Hanson in 1887.

The present chaplain is the Rev. Gerard M. Mason, who succeded Mr. Rudge in March 1886. Mr. Rudge had served thirty years with the schools, and his retirement was deeply regretted. Mr. Foster, of Fernside, Witley, added a gymnasium in 1887 as a Jubilee offering and memento, and the Governors a splendid playroom of large dimensions and a carpenter's shop, to enable the boys to be further instructed in manual and instructive pursuits. In the last reports of examination by Mr. Waddington, most satisfactory results are recorded as to the condition of the inmates of both schools. The needlework of the girls is noticed particularly; also their examination in religious subjects by the diocesan inspector. The drilling of the boys, their excellent play in the band, and their general appearance never appeared better.

On the day in July for the annual examination, many of the Governors and their friends find their way down to Witley, and one can testify to their intense pleasure in witnessing the busy scene presented by 240 boys, in and out of school; the prize-giving, the out-of-door exercises, &c.; and the most satisfactory thing of all is the reflection that these children have for the most part been taken from bad influences, reared for years under careful discipline, with simple, plain, and good education, founded upon the tenets and principles of the Church of England, so as to fit them to battle with the world in after life like good citizens and soldiers in life's contest.

KING EDWARD'S SCHOOL FOR GIRLS, LATE HOUSE OF OCCUPATION.
From a Drawing by ALLAN BARRAUD, 1883.

CHAPTER XII.

ENDOWMENTS OF THE HOSPITAL.

THE expense of the first establishment of Bridewell was not defrayed by voluntary contributions, but by a compulsory assessment of the City Companies. The Hospital, however, did not share very largely in the bounty of the citizens.

In 1602 the fines and forfeitures accruing from the constables and others, for not punishing rogues and beggars, according to the statute 39 Elizabeth, were handed over for the maintenance of the House, while another privilege belonging to it was the right of collecting rags and marrow-bones.

St. Thomas's Hospital appears to have appropriated to its use, the lands granted by King Edward VI., the precinct of the ancient palace being all that was set apart for Bridewell, and for years it was a matter of considerable anxiety to the Court of Aldermen how the latter should be maintained.

St. Thomas's paid £200 a year quarterly to Bridewell, with slight intermission, from 1589 to 1670; but after falling into arrear, the payment finally lapsed in 1684.

H

Considerable benefactions from time to time have
been made by way of gift or grant, but the present
condition of the Hospital stands somewhat thus in
the possession of real estates and annuities:—The
former consists of houses in London and Middlesex, a farm in Oxfordshire, and an estate at Wapping; and it shares with Bethlehem the rents of a
farm in Kent.

Since the prison was pulled down and demolished,
and the leases in New Bridge Street fell in, the old
Bridewell precinct has been materially altered.   A
good road now passes through the back, and the
whole available space is covered with large houses and
offices, notably Messrs. Spicer and Son's paper warehouses, and the Royal Hotel, whose proprietor is Mr.
Alderman De Keyser, Lord Mayor of London, 1887–88.

No. 14 in New Bridge Street is easily recognised
as Bridewell Royal Hospital, and contains within its
area the treasurer's house, the beadle's lodge, the
offices and hall, and the receiver's house, together
with the cells for recalcitrant apprentices who may be
sent for solitary confinement by the City Chamberlain.

The lands at Wapping and the Rectory at North-
leigh (Oxon), with the appurtenances as described in
the grant of Henry VIII., to be held *in capite* by
the service of the fortieth part of a knight's fee, was,
for the consideration of a fine of £184, 5s., confirmed
by Queen Elizabeth, April 12, 1600.

In 1759, on the enclosure of the parish of North-leigh, the Commissioners awarded to the Governors 323 acres and 16 perches, and the advowson of the Vicarage was vested in the Crown.

The estate at Wapping, formerly a marsh, is now covered with wharves, warehouses, manufactories, and a few private dwellings, with a river frontage granted by letters patent, Charles II., 8th April 1676.

The incidents connected with this estate are fully related by Mr. Martin in his report of 1837, referred to in Chapter X.

The north side or entrance of the Old Thames Tunnel is situated on the Wapping estate, and it has been affirmed, on somewhat doubtful authority, that the principal grant which Charles II. appears to have made at Wapping was of some ground near the Thames, as compensation for land lying near "Bridewell Dock," and on which he had pro-hibited any new buildings being erected, after the Great Fire, to replace those which had been destroyed.

Bridewell itself is situate in the ward of Farringdon Without, a ward that occupies one-fifth of the whole City, and is conspicuous for the enormous interests it contains. There are, besides those of the Temple, banking, commercial, and market interests, and it is also the centre of the great printing and publishing trades, from which emanate, besides general litera-ture, over 130 newspapers, daily and weekly, for the education and amusement of the community at large. The present Lord Mayor, Polydore De Keyser, Esq., is the Alderman of the ward, and the deputy, Mr. Walter, has resided in the ward seventy years, and represented it in the Court of Common Council for forty-four years.

The pecuniary bequests of Locke, Fowke, and Palmer, for helping apprentices of good character, were administered up to a recent date, and when the new scheme was adopted, the interest was absorbed in the general fund.

The management or government of the Hospital, according to the provisions of the Act 22 George III., c. 77, is vested in a president and a treasurer, the Court of Aldermen, twelve Common Councilmen, elected as the Act directs, and an unlimited number of Governors, who have presented the Hospital funds with a benefaction of fifty guineas.

The revenues of the Hospital being considerably diminished by the present depression in trade and agriculture, the treasurer welcomes these benefactions with considerable interest.

The treasurer has the use of a furnished house at Bridewell, but he receives no salary, and no part of the moneys of the Hospital passes through his hands.

All Governors have equal voice in any election.

Various officers are appointed to do the necessary business of the Hospital or Institution, as it should be more correctly called, and the same administers

TREASURER'S RESIDENCE, BRIDEWELL,
14 NEW BRIDGE STREET, BLACKFRIARS, E.C.

to the duties involved in Bethlehem Hospital, which is incorporated with that of Bridewell.

These include a surveyor, solicitor, agents for the Lincolnshire and Kent estates, steward, clerk and receiver, resident physician, chaplain and superintendent, with their necessary subordinate officers.

Each Governor, on being admitted, receives the following charge, which was written by Bishop Atterbury :—

"THE CHARGE TO EVERY GOVERNOR ON HIS ADMISSION.

" *Given in the presence of the President or Treasurer, and other Governors, assembled in Court.*

" SIR,—You have been elected, and are come to be admitted, a Governor of the Royal Hospitals of Bridewell and Bethlem, a station of great honour and trust, which will afford you many opportunities of promoting the glory of God and the welfare of your fellow-creatures ; for in these Hospitals a provision is made for [*employing and correcting idle, vagrant, and disorderly persons, and*[1]] educating poor children in honest trades, and also for maintaining and curing needy and deplorable lunatics.

" The distribution of the revenues designed by

---

1 These words, which are in the original, are at the present time usually omitted.

royal bounty and many charitable persons for those truly noble and excellent purposes, is now about to be committed to your care; and you are hereby solemnly required and earnestly requested to discharge your duty in this behalf with such conscientious regard, that you may appear with joy at the judgment-seat of Christ, when a particular account will be taken of all the offices of charity in which we have abounded towards our poor brethren, and a peculiar reward conferred on those who have with fidelity and zeal performed them.

"In confidence that you will diligently attend to this good work, you are now admitted a Governor of the Hospitals of Bridewell and Bethlem."

Atterbury became preacher of Bridewell Hospital and Minister of Bridewell Precinct in 1693. He was subsequently appointed Dean of Carlisle. In 1713 he resigned his position at Bridewell on being promoted by Queen Anne to the Deanery of Christ Church, Oxford. In 1714 he was elected a Governor of the Royal Hospitals of Bridewell and Bethlehem. A few months later he was made Bishop of Rochester and Dean of Westminster. On suspicion of being implicated in a plot in favour of Charles Edward, the Pretender, he was imprisoned in 1722, and by Act of Parliament was deprived of all his dignities and offices and condemned to perpetual exile. He died in Paris in 1732, but his remains are laid in Westminster Abbey.

Stow writes of the chapel at Bridewell that it "was enlarged and beautified at the proper cost and charge of the Governors and inhabitants of this precinct in the year of our Lord 1620, Sir Thomas Middleton being then president, and Master Thomas Johnson treasurer of this Hospital.

"This enlargement was by the taking in of a large room, that, before the date above - named, joyned upon the head of the chappell. This ground adding to the length of it (all the full breadth going with it) 24 foote and better.

"This room thus taken in, trimmed, beautified, and consecrated is now a beautiful chappell, it being before a room unfit, vast, rude, and unsightly, though then in the use deserving a fair commendation.

"For then that ground that is now a church to the prisoners of the house was a chapell, into which every Sabbath (through a bye or backward passage) they were brought from their severall lodgings to heare divine service.

"So that then and now in that worthy use, and this worthy alteration and beauty we may see the pious and religious care of these worthy right worshipfull Governours continually employed and applyed to things of this excellent nature."

Among the records now at the Boys' School at

Witley are the following notes, some of passing interest.

October 4, 1693.—The Rev. Francis Atterbury, elected preacher of Bridewell Hospital and minister of Bridewell Precinct.

June 15, 1713.—He resigned, upon his appointment to the Deanery of Christchurch.

February 26, 1714.—He was elected a Governor of the Royal Hospitals of Bridewell and Bethlehem.

June 26, 1713.—The Rev. Thomas Yalden, D.D., elected minister and preacher.

October 27, 1736.—The Rev. William Gibbon, M.A., elected.

February 8, 1758.—The Rev. Moses Wright, M.A., elected.

February 16, 1770.—Elected a Governor.

December 8, 1774.—The Rev. Thomas Bowen, M.A., elected reader.

January 29, 1784.—Elected a Governor.

January 29, 1795.—The Rev. Moses Wright's decease reported. On his death the offices of reader and preacher were by order of Court consolidated, and the same day the Rev. Thomas Bowen was elected chaplain of Bridewell Hospital and minister of Bridewell Precinct.

January 16, 1800.—Rev. Thomas Bowen died, *ætatis* 51. Following wise and good men, he lived in the faithful discharge of the sacred duties of his

FRANCIS ATTERBURY, BISHOP OF ROCHESTER.
From an Old Print.

office, and manifested his zeal for the benefit of these Hospitals by his able and useful publications.

February 19, 1800.—The Rev. Henry Budd, B.A., elected chaplain.

March 24, 1831.—Resigned. Elected a Governor, April 11, 1832.

April 12, 1831.—The Rev. Robert Monro, M.A., elected chaplain, &c. ; resigned March 26, 1849.

April 30, 1849.—The Rev. Frederick Poynder, M.A., elected chaplain, &c. ; resigned Michaelmas 1858.

1833.—Thomas E. Garrett, B.D., appointed chaplain to House of Occupation ; superannuated 1856.

May 1856.—Edward Rudge, LL.B., elected chaplain ; elected superintendent, King Edward Boys' School, in November ; removed to Witley, March 1866 ; superannuated January 1886 on resignation, and left March 1886.

When the Boys' School was removed to Witley, an assistant chaplain was appointed for the Girls' School, the present holder of the office being the Rev. B. West.

On St. Matthew's Day, 21st September in each year, the Lord Mayor and the Sheriffs and Aldermen, according to custom, go in state to attend divine service at Christ Church, Newgate Street. The Blue Coat Boys of Christ's Hospital attend the service, after which an adjournment is made to the Hospital, and the lists of the Governors of all the Royal Hospitals are presented to the Lord Mayor.

It was also the custom on Easter Monday, altered a year or two since to the Tuesday, for a sermon, termed the "Spital Sermon," to be preached by one of the Bishops in Christ Church, Newgate Street, before the Lord Mayor, the Corporation, and the Governors of the Royal Hospitals.

The Blue Coat Boys also attend, after being regaled at the Mansion House with a bun, a glass of wine, and a "tip" from the Lord Mayor. They wear on their coats a piece of silk embroidered with the legend "He is risen," referring of course to the great event commemorated on Easter Day.

The Lord Mayor, on a convenient evening, entertains numerous guests, and the toast of the Royal Hospitals is the great toast of the evening.

This Spital sermon derives its name from the Priory and Hospital of Our Blessed Lady St. Mary Spital, which was situated on the east side of Bishopsgate Street, with fields in the rear, which now form the suburb called Spitalfields.

Hard by this hospital, founded in 1197, was a large churchyard with a pulpit cross in it, from whence it was an ancient custom on Easter Monday, Tuesday, and Wednesday, for sermons to be preached on the resurrection before the Lord Mayor, Aldermen, Sheriffs, and others, who sat in a house of two storeys for that purpose, the Bishop of London and other prelates being above them.

In 1594 the pulpit was taken down and a new one set up, and a large house erected for the Governors and children of Christ's Hospital to occupy.

In April 1550 Queen Elizabeth came in great state from St. Mary's, Spital, attended by a thousand men in harness, with shirts of mail, and corslets, and pikes; and ten great pieces of ordnance, were carried through London into the court, with drums, flutes, and trumpets sounding, morris-dancers, and two white bears in a cart.

On Easter Monday, 1617, James I. having gone to Scotland, the Archbishop of Canterbury, the Lord Keeper Bacon, the Bishop of London, and certain other lords of the court and privy councillors, attended the Spital sermon with Sir J. Lemman, the Lord Mayor, and Aldermen, and afterwards rode home and dined with the Lord Mayor in his house near Billingsgate.

The Hospital was dissolved under Henry VIII.; and the pulpit broken down during the troubles of Charles I.

After the Restoration the sermons denominated "Spital" were preached at St. Bride's, Fleet Street, on the three usual days.

A writer of the last century speaks of a room crammed as full of company as St. Bride's Church upon the singing a Spital psalm at Easter or an anthem on Cecilia's Day. For many years past the sermons have been preached at Christ Church, Newgate Street.

The following is a list of the Presidents and Treasurers of Bridewell and Bethlehem Hospitals from 1557 to the present time :—

Presidents.	Year.	Remarks.
Sir Rowland Hill, Knt......	1557	
Sir Wm. Garrett, Knt.......	1558	
Sir Rowland Hill, Knt......	1559	...
Sir Roland Haywood, Knt.	1561	...
Edward Gilbart...............	1563	...
Sir Wm. Chestie, Knt.......	1564	...
Sir Jno. White, Knt.........	1568	Grocer, and Lord Mayor, 1563. No feast on account of plague; Thames frozen over.
Sir Alex. Avenon, Knt......	1573	Ironmonger : eight times Master.
Sir Lionel Duckett, Knt....	1580	Mercer. President from 1569 to 1573 and from 1580–1586.
Sir Wm. Row *or* Rowe, Knt.	1592	Ironmonger : five times Master.
Sir Wm. Webbe...............	1594	Salter.
Sir Stephen Slaney..........	1599	Skinner. President of Christ's Hospital, 1602.
Sir Wm. Ryder...............	1600	...
Sir Leonard Halliday........	1605	Merchant Taylor.
Sir Thos. Bennett	1606	..
Sir Thos. Middleton..........	1613	Grocer. His younger brother, Sir Hugh, projected the New River, opened with great splendour on the day Sir Thomas was Lord Mayor.
Sir Roland Hayter............	1631	...
George Whitman..............	1631	...
Sir Jno. Wollaston, Knt.....	1643	...
Christopher Pack.............	1649	...
Sir Richd. Brown, Bart......	1661	Clothmaker.
Sir Jas. Smith, Knt..........	1668	...

Presidents.	Year.	Remarks.
Sir Wm. Turner, Knt.[1]	1669	Merchant Taylor.
Sir Robert Jeffries, Knt.	1689	...
Sir Wm. Turner, Knt.	1690	...
Sir Robert Jeffries, Knt.	1693	...
Sir Samuel Dashwood, Knt.	1703	...
Sir Thos. Rawlingson, Knt.	1705	...
Sir Wm. Withers, Knt.	1708	...
Sir Samuel Garrard, Bart.	1721	...
Humphrey Parsons	1725	...
Robt. Willemott	1741	...
Wm. Benn	1746	Goldsmith.
Sir Richd. Glynn, Knt.	1755	...
Sir Walter Rawlinson	1773	...
Brackley Hennett	1777	...
Brass Crosby	1782	...
Sir Jas. Saunderson, Knt.	1789	...
Sir R. Carr Glynn, Bart.	1793	...
Sir Peter Laurie, Knt.	1833	Saddler.
W. T. Copeland	1861	Goldsmith.
J. E. Johnson	1868	...
Sir Jas. C. Lawrence, Bart.	1868	Fishmonger.

TREASURERS.

Treasurers.	Year.	Treasurers.	Year.
John Haskerne	1557	Henry Johnson	1618
Will. Box	1560	John Whitwell	1626
John Buckland	1563	John Withers	1631
Edwd. Mabbe	1572	John Rawlins	1637
Thos. Gardiner	1576	Robt. Edwards	1641
John Ladington	1580	Henry Isaacson	1642
Harry Warfield	1581	Gawin Gethin	1654
Gabriel Newman	1592	Benj. Ducane	1672
Thos. Caldwell	1599	Robt. Baker	1683
Richd. Arnold	1600	Hy. Ducane	1688
Richd. Wyat	1602	Dan. Baker	1693
John Polland	1603	James Gardiner	1700

[1] There is in Merchant Taylors' Hall an excellent portrait, by G. Kneller, of this gentleman.

Treasurers.	Year.	Treasurers.	Year.
Sir Jas. Cass, Knt......	1709	Nathl. Thomas..........	1775
John Tayler..............	1714	Richd. Clarke [1]	1781
Robt. Alsop..............	1729	Ralph Price..............	1836
Robt. Bishop....	1737	J. E. Johnson [2]	1848
Edwd. Holloway........	1739	Sir Chas. Hood, Knt.[3]	1868
Robt. Alsop..............	1750	John Baggallay [4]	1870
John Wallington........	1755	A. J. Copeland..........	1885
Wm. Kinlisioe	1768		

[1] Fifty-five years Treasurer.   The bust of Mr. Clarke, on the staircase at Bridewell, presented by Mr. Hardwick in 1837, bears the following inscription :—

SACRED

TO THE MEMORY OF

RICHARD CLARK, Esq.

WHO DIED 16 JANUARY 1831,

IN HIS 92ND YEAR.

———

He filled the office of Treasurer
of Bridewell and Bethlem Hospitals
for the space of half a century
with the highest honour and integrity,
and with great advantage to those Institutions.

He was the legal pupil of Sir John Hawkins,
the personal friend of Dr. Samuel Johnson,
and the respected associate
of many of the literary and most esteemed characters of the
last and present age.

He was endowed with a mind of the most amiable qualities,
his manners were eminently attractive and engaging.
and he enjoyed to the latest period of a protracted life
the affection and attachment
of all with whom he was connected,
either in its private relations
or its public duties.

———

[2] Afterwards President.
[3] Previously Head-Physician at Bethlehem.   [4] Retired.

Among the numerous addresses that were presented to her present Majesty in her Jubilee year,
1887, one was included from the Royal Hospitals,
drawn up by Mr. Cross, of St. Bartholomew's, and
signed by the four Presidents and Treasurers, as
follows :—

"*To the Queen's Most Excellent Majesty.*

" We, your Majesty's loyal subjects, the Presidents,
Treasurers, and Governors of the Royal Hospitals
of St. Bartholomew, Christ, Bridewell, and Bethlem,
and St. Thomas the Apostle, of the City of London,
humbly beg leave to be permitted to offer to your
Majesty our most dutiful and sincere congratulations
on the completion of the fiftieth year of your most
propitious and eventful reign.

" Dedicated in bygone ages to works of piety and
deeds of love, the care and healing of the sick
poor, and the nurture and instruction of the young,
and incorporated and endowed by your Majesty's
illustrious predecessors King Henry the Eighth and
King Edward the Sixth, the Royal Hospitals have
conferred, during many generations, incalculable
benefits on the several objects of their bounty.

"To your Majesty each one of the Royal
Hospitals is deeply indebted, as we gladly acknow-
ledge, for many acts of gracious favour and con-
descension ; and we rejoice that of one of them,

Christ's Hospital, your Majesty's name has been for many years enrolled as a Governor.

"It is to us a matter of the most lively satisfaction that that true spirit of philanthropy which led to the founding of the Royal Hospitals is ever growing amongst your people, and is evidenced not only by the continued and increased efficiency of these our ancient foundations, but by the establishment and maintenance of numerous other charitable institutions designed to meet the always present wants of the sick, the sore, and the needy,—a fact which we cannot but recognise as in great measure attributable to your Majesty's noble example of large-hearted and womanly sympathy with the sorrows and afflictions of all classes of your subjects.

"With feelings of sincerely attached loyalty to your Majesty's person and throne, and with gratitude for your life's devotion to the well-being and prosperity of your people, we humbly and earnestly pray that it may please Almighty God to continue to guide and guard your Majesty with His protecting hand, and to preserve your life for many years in health, peace, and happiness."

In the court-room of Bridewell Hospital there are some most interesting historical paintings and portraits. The following is a list :—

Presentation of the Charter of the three Hospitals of Christ, Bridewell, and St. Thomas by King Edward the Sixth. *This picture is of great historical value, and until comparatively recently was attributed to Holbein.*

Portrait of King Charles the Second.   By Lely.

,,   King James the Second.   By Lely.

,,   Sir William Turner, President 1669.   By Mrs. Beale.

,,   Sir Robert Geffery, President 1693.   By Kneller.

,,   Sir Thomas Rawlinson, President 1705.   By Kneller.

,,   Sir Samuel Garrard, President 1720.   Unknown.

,,   William Benn, Esq., President 1746.   By Hudson.

,,   Sir Richard Glyn, President 1755.   By Zoffanni.

,,   Sir James Sanderson, President 1793.   By Dupont.

,,   Sir Richard Carr Glyn, President 1798.   By Hoppner.

,,   Sir Peter Laurie, President 1833.   By Frazer.

,,   King George the Third.   *Copy after Romney.*

,,   Queen Charlotte.   *Copy after Romney.*

A portrait unknown.   By Lanskræn.

Portrait of Richard Clark, Esq.. Treasurer 1781.   By Lady Bell.

,,   Ralph'Price, Esq., Treasurer 1836.   By R. P. Knight, R.A.

,,   John Edward Johnson, Esq., Treasurer 1848.[1]   By Tweedie.

,,   William Taylor Copeland, Esq., President 1861.   By Tweedie.

---

[1] Subsequently President.

CHAPTER XIII.

EXTRACTS FROM "REMEMBRANCIA."

THE following are curious, as having reference to Bridewell, whether as hospital or prison. They are extracts from the archives of the City of London, A.D. 1579–1664. There is a letter, 14th January 1579, from the Lord Mayor to Sir George Carey,[1] bringing to his notice, the complaint made against his servant Lucas, for using abusive and threatening words towards Robert Winch, Treasurer of Bridewell, and requesting him to take steps to prevent the repetition of such conduct; likewise informing him that his servant Gold, who had been permitted to lodge in Bridewell, had so conducted himself against the City that he would not be suffered to remain there.

The Court of Aldermen had been informed of his intention, to make a request for a part of that house

---

[1] Sir George Carey, eldest son of Henry, Lord Hunsden, cousin of Queen Elizabeth ; knighted 1571 ; succeeded to the title as second Lord Hunsden on the death of his father in 1596 ; made Lord Chancellor, March 1597 ; Lord Chamberlain of the Household, 1598 ; K.G., April 22, 1592 ; died, 1603.

for himself. It was the intention of the City, to employ the place for the storage of corn and other such public uses.

A LETTER (dated Somerset House, 15th January 1579) from Sir George Carey to the Lord Mayor and Court of Aldermen in reply, denying the imputation made against his servants, and alleging that the Treasurer was a person unworthy of credit. It had not been his intention to request a part of Bridewell for himself, but for a friend who had intended to pay for the same.

A LETTER from the Lords of the Council, 6th April 1582, to the Lord Mayor, stating that they had been lately informed that a gentlewoman of good birth and alliance, Mrs. Moody, had, upon some suspicion of ill-behaviour, been committed to the Compter, and from thence removed to Bridewell.

Some of her friends had caused her to be rescued by the way, in which attempt, one of the beadles was casually slain.

The Council requested an inquiry into the whole case to be made, and if it should appear that she had not been a party to the officer's death, she should be set at liberty.

A LETTER from the Lord Mayor to the Lords of the Council in reply.

They had been misinformed of her faults, whatever had been stated in her excuse, touching her privity to the rescue, whereupon the murder or manslaughter ensued ; the plea "that she would not know of it by reason of her close imprisonment," had been stated rather to move their compassion, than for matter of truth.

The poor woman, the wife of the man that had been killed, having lost her husband and the means of sustenance, desired justice against this woman.

Before the receipt of their letter steps had been taken to release her, upon reasonable security being given for her appearance to answer the charge; her enlargement, however, had been stayed until the Council's further pleasure had been ascertained.

A LETTER from the Lord Mayor to the Lord Chamberlain, dated 8th July 1614, detailing the steps taken by him since his appointment for reforming, what he found out of order in the City.

Firstly, He had freed the streets of a swarm of loose and idle vagrants, providing for the relief of such as were not able to get their living, and keeping them at work in " Bridewell ;" not punishing any for begging, but setting them on work, which was worse than death to them.

Secondly, He had informed himself, by means of spies, of many lewd houses, and had gone himself disguised to divers of them, and finding these nurseries of villainy, had punished them according to their deserts, some by carting and whipping, and many by banishment.

Thirdly, Finding the gaol pestered with prisoners, and their bane to take root and be beginning at ale-houses, and much mischief to be there plotted, with great waste of corn in brewing heady strong beer. "Many consuming all their time and means sucking that sweet poison," he had taken an exact survey of all victualling houses, amounting to above 40,000 barrels ; he had thought it high time to abridge their number, and limit them by bonds as to the quality of beer they should use, and as to what orders they should observe, whereby the price of corn and malt had greatly fallen.

Fourthly, The bakers and brewers had been drawn within bounds, so that if the course continued, men might have what they paid for, viz., weight and measure.

He had also endeavoured to keep the Sabbath-day holy, for which he had been greatly maligned.

Fifthly, If what he had done were well taken, he would proceed further, viz., to deal with thieving brokers or broggers, who were the receivers of all stolen goods. And, lastly, the inmates and divided houses would require before summer to be discharged of all superfluities for avoiding infection.

A LETTER from the Court of Aldermen to the Lords of the Council, 12th February 1590, acknowledging their letter on behalf of Ferdinando Richardson and Mr. Richard Tothill, for the renewal of the estate of the said Tothill in certain tenements pertaining to the Hospital of Bridewell.

They had the Governors of the Hospital before them, and commended the same to their considera-tion, and had since received their reply, from which it appeared, that on account of the extraordinary charges of the charity, the Governors had already granted a reversion of Tothill's lease to the several tenants, in consideration of certain charges incurred by them, in repairing the tenements for the benefit of the Hospital.

They regretted that, for the above reasons, they were unable to comply with the Council's request.

A LETTER from the Lords of the Council, 31st December 1594, to the Lord Mayor, the Archbishop of Canterbury (Whitgift), and others, concerning the Commission given under the Great Seal to inquire into the manner in which the lands in that county (not named) belonging to the Hospital were employed or abused, and to see if any provision could be made for the sustentation and comfort of maimed soldiers, who were not sufficiently provided for by the statutes.

A LETTER, about 1600, from the Lord Mayor to Mr. Cooke (Sir Edward Coke), Attorney-General, beseech-ing his good offices in behalf of the City, in the settle-ment of the question referred to him and the Recorder, touching the lands and tenements in question between the Hospitals of Bridewell and Bethlehem and Mr. William Tothill.

A LETTER from the Lord Mayor to the Earl of Dorset,[1] 23d September 1608, touching a parcel of ground lying on the west part of Bridewell Hospital, belonging to the President and Governors, which of late had been enclosed by his father without consent of the Governors, and praying that the same might be restored.

A LETTER dated 28th March 1611 from the Lord Mayor to Lord Woolton,[2] in reply to his application on behalf of Ann Tisdale for a lease of her dwelling in a part of Bridewell, stating that the President and Governors desiring him to acquaint his lordship that, for the better government of the said Hospital and the *relief of poor fatherless children there*, they had agreed that none should inhabit or hold any part of it by lease, except officers of the place and such artificers as, having fitting trades, would be bound to take poor children as apprentices; but that they had, in consideration of her father and grandfather having

---

[1] The old mansion and manor of Salisbury Court, *alias* Sackville Place, *alias* Dorset House, was confirmed to Richard, Earl of Dorset, March 25, 1611, the family having held it some years previously.

[2] Thomas, second Lord Woolton, of Marley, Kent, succeeded to the title, 1604; Treasurer of the Household, 1616–18; died 1630, when the title became extinct.

been dwellers there, and of their expenditure on the premises, permitted her and her husband (who was only a tailor, and not bound or able to take and bring up poor children as apprentices) to remain as tenants at will.

A LETTER from Lord Verulam, Lord Chancellor, to the Lord Mayor, dated York House, 3d December 1619.

The French Ambassador, Comte de Tilliers, had desired that the punishment to be inflicted upon certain persons, committed to Bridewell for their insolent and outrageous assault upon him and his people, might be remitted; upon which the Lords of the Council had thought fit, they should be discharged without further punishment, but that first they should be carried by their keeper to the Ambassador, if he would see them, otherwise he was to be informed that they were sent to ask his forgiveness on their knees, and then be set at liberty by his Grace.

This arose out of an occurrence on 28th October 1619, when a tumultuous assemblage took place before the house of the French Ambassador, resulting from a quarrel between his boy and a carman, in which his servants and others, passers-by, took part. A constable who went to appease them, being taken into the Ambassador's house, a report circulated that he was slain there, and much uproar arose, which was stilled by his reappearance, when the people dispersed.

A LETTER from Sir Thomas Smyth to the Lord Mayor recites the following, under date 13th Jan. 1618 :—

That "the King to Sir Thomas Smyth, states that the Court had lately been troubled with divers idle young people, who, though twice punished, still continued to follow the same ; having no other course to clear the Court from them, had thought fit to send them to him, that at the next opportunity they might be sent to Virginia to work there," and remarking that some of these persons had already been brought by the King's command from Newmarket to London, and others were coming.

The Company of Virginians had no ship ready to sail, and no means to employ them or place to detain them in, and he requested the Lord Mayor to authorise their detention and employment in Bridewell, until the next ship should depart for Virginia.

The Lord Mayor received an intimation from the Council, informing him that all the ills and plagues affecting the city were caused through the number of poor swarming about the streets, and recommending the Corporation to subscribe with the Companies and the several wards, and so to raise a fund to ship out these persons to Virginia ; and he issued his precept to the several Companies for the purpose, March 27, 1609.

On April 29 the Merchant Taylors' Company determined to subscribe £200, and the members of the Company advanced £300 more.

The Ironmongers advanced £150. £1800 was raised in the City for the purpose of founding this plantation.

A broadside was issued in 1610 by the Council of Virginia touching this plantation ; another stating that a good fleet of ships under the conduct of Sir Thomas Gates and Sir Thomas Dale, Knights, would soon be ready to sail, and directing good artificers and others desirous of joining, to repair to the house of Sir Thomas Smyth, in Philpot Lane, before the end of January 1612.

A broadside was also issued in February 1621, giving the numbers of the ships and people (one being the *Mayflower*) sent out from August 1620 to February 1621.

A LETTER from William, Lord Beauchamp (Netley, 22d July 1620), to the Lord Mayor and Court of Aldermen, with respect to the City's wall adjoining his house and garden in Blackfriars (in margin "over against Bridewell"), which was in so great ruin that if not speedily taken in hand it could not be restored before winter.

He was advised by counsel that he could not contribute thereto without prejudice to himself and posterity, but he pledged his honour, so tender a care had he for the City's right, that if anything were justly proved, he would not be unready to give due satisfaction.

When patents were granted by King James I. for the issue of farthing tokens, a proclamation was issued in 1633, by which it was ordered that counterfeiters of these tokens upon conviction should be fined £100, be set in the pillory in Cheapside, and from thence whipped through the streets to Old Bridewell, and there kept to work; and when enlarged, should find sureties for their good behaviour.

A very curious reminiscence of Bridewell is found as follows in the reign of Queen Elizabeth, from weekly reports by Mr. Fletewoode, Recorder of London, to Lord Burghley :—

" My singular good Lord, uppon Thursdaye at even, her Majestie in her coache nere Islington taking of the air, Her Highness was environed with a number of roogs. One Mr. Stone a footman cam in all haste to my Lord Maior, and after to me, and told us of the same. I dyd the same nyght send warrants out into the sayd quarters and into Westminster and the Duchie, and in the morning I went abroad myselff and tooke that day lxxiiij roogs, whereof some were blynde and yet great usurers and very rich ; and the same daye towards nyght I sent Mr. Harrys and Mr. Smithe, Governors of Bridewell, and took all the names of the roogs, and then sent theym from the Sessions Hall into Bridewell, where they re-

mayned that nyght. Uppon Twelff daye in the fore-
noone, the Master of the Rolls, myselff and others,
receyved a charge before my Lords of the Counsell
as touching roogs and masterless men and to have
pryvie searche. The same daye at after dyner (for I
dyned at the Rolls) I met the Governors of Bridewell, and so that afterwards wee examined all the
seyd roogs and gave them substantial payment. And
the stronger we bestowed on the mylle and the
lighter; the rest were dismyssed with the promise
of a double paye if we met with them agayne. Uppon
Soundaye being crastine of the Twelffth daye, I dyned
with Mr. Deane of Westminster, when I conferred
with hym touchinge West'. and the Duchie. And
then I took order for Southwarke, Lambeth, and
Newyngton, from whence I receyved a shoal of xi.
roogs, men and women and above. I bestowed
them in Bridewell. I dyd the same afternoone
peruse Pools (St. Paul's), when I tooke about xxii.
cloked roogs that there used to kepe standing. I
placed theym also in Bridewell. The next morning
being Mundaye the Master of the Rolls and the reste,
tooke order with the constables for a privie searche
agaynst Thursdays at nyght, and to have the offenders
brought to the Sessions Hall uppon Fridaye in the
mornyng, where wee the Justices shold meete. And
agaynst the same tyme my Lord Maior and I did
the lyke in London and Southwarke. The same

afternoone the master of Bridewell and I mett and after every man had been examined, eche one receyed his payment according to his deserts: at which tyme the strongest were put to worke, and the others dismissed into their countries: the same day the Master of the Savoye was with us, and said he was sworn to lodge, 'Clandicantes, egrotantes, et perigrinantes,' and the next morning I sent the constables of the Duchie to the Hospitall, and they brought unto me at Bridewell vj tall fellows, that were draymen unto brewers and were neither 'clandicantes, egrotantes, nor perigrinantes.' The constables, if they might have had theyre owne will, would have brought in many moor. The Master did write a very cartese letter unto us to produce theym, and although he wrote charitably unto us, yet were they all soundly paydd and sent home to their masters. All Tuesdaye, Wednesdaye, and Thursdaye, there came in numbers of roogs : they were rewarded all according to theyr deserts. Uppon Fridaye morninge, at the Justice Hall, there were brought in a 100 lewd people taken in the private searche. The Master of Bridewell receyed theym, and immediatly gave them punishment. The Saturdaye after causes of consciense herd by my Lord Maior and me, I dyned and went to Polls (St. Paul's), and in other places, as well withine the libertes as elsewhere. I founde not one rooge styrryng. Ermongst all these thyngs I dyd note

that we had not of London, Westminster, nor Southwarke, nor yett Middlesex nor Surrey above twelve,
and those we have taken order for. The residue for
the most part were of Wales, Salop, Cester, Somerset,
Barks, Oxforde, and Essex, and that few or none of
them belonged about London above iij. or iiij. months.
I'd note also that wee mett not agayne with many in
all our searches, that had receyed punishment. The
chieff nurseries of all these evill people is the Savoye
and the brick kilns near Islington. As for these
brick kilns we'll take such order that they shalle be
reformed, and I trust by your good Lordships help,
the Savoye shall be amended as surelie.

"As by experience I fynd it the same place as it
is used is not converted to a good use or purpose,
and this shall suffice for roogs.   W. E. C.

"The Savoy was the great nest of these roogs, and
in consequence of objection being made to it on this
account, the Master comes to the writer to excuse
himself.   Mirror, xviii., p. 337."

CHAPTER XIV.

JOTTINGS FROM OLD NEWSPAPERS.

THE following jottings from old newspapers will be found curious and interesting in their references to Old Bridewell :—

December 1719.—Edmund Thomas at the "Old Brunswick Mum and Spruce Beerhouse" against Bridewell Bridge, Fleet Ditch, sells right Brunswick Mum and Spruce Beer wholesale and retail. Note, he hath a large quantity of new spruce just arrived, and is the only person in London that deals in these two commodities and nothing else.

October 1719.—A Convenient House and Coal Wharf in Bridewell Dock well accustomed, to be disposed of, with the advantage of the Trade, the person wishing to leave off his business.

Enquire at the Golden Key, between Fleet Bridge and Bridewell.

May 1720.—Whereas in the late fire on Sunday night, the 1st inst., in Bridewell Precinct, several goods and things of value are missing, viz., a gold watch and chain marked on the inside case Gibraltar, made by one Morley, and plate, herein, &c.   These are to give notice, that if any person will bring them or any part thereof, or give information where the same may be had, they shall be well rewarded by Thos. Arnitt at C. Bateman's, Bookseller, in Paternoster Row, the corner house next Warwick Lane.

May 19, 1720.—Whereas it is incidentally reported that Mr. Theophilus Arnitt, woodmonger, at the lower end of Bridewell Dock (who has had a very great loss by fire), has left off his trade.   This is to certify his friends that he still continues the woodmonger trade as before, at the same place.

August 1720.—John Howe, Esq., of Gloucestershire, is chosen one of the Governors of the Hospital of Bridewell and Bethlem, having given £100 to the latter for the use of such distracted persons as are declared incurable.

March 1722.—On Saturday last at 12 of the clock a fire broke out in the New River Company's House at the lower end of Bridewell Dock, which might have been of dangerous consequence had it not been timely extinguished by the diligence of the watermen belonging to the Sun Fire Office.

July 1723.—Mr. Hart, formerly partner to Mr. Taylor, a governor of Bridewell, was well in health on Tuesday night, but found dead in his bed the next morning.

Being a gentleman of an extraordinary good character, his death is much lamented by all who knew him.

September 4, 1723.—Yesterday evening, one Bird, a watchmaker, aged above 20, living in Bridewell Precinct, cut his own throat, but not doing it effectually, hanged himself afterwards on the bannister of the staircase.

August 6, 1730.—Mr. Alderman Parsons has wrote to Mr. Alderman Child from France, to desire he will hold the annual Court Thursday at Bridewell, where a fine entertainment will be prepared at the expense of twelve stewards, the said Alderman having put off his return to England for some time.

1740.—On Saturday last the son of Mr. Woolton, a glazier in Bridewell Precinct, a youth of about seven years of age, playing with his companions in a lighter in Fleet Ditch, before his father's door, unfortunately fell overboard and was drowned, which frightful accident so terrified his playmates that, with hastening out of the lighter, they had all like to have shared the same fate.

1740.—Thursday a Court was held at Bridewell Hospital for the election of two art-masters. The candidates were—

    Mr. Thomas Keil, engine-weaver.
    Mr. Wm. Dell, weaver.

                      K

Mr. Wm. Simmonds, fanstick-maker.

Mr. Cornelius Mortier, velvet-weaver.

Mr. John Benchan, weaver.

The Weavers' Company in great numbers, masters and journeymen, attended a petition against the admission of Mr. Thomas Keil, and after both sides heard, and a ballot, the numbers stood thus :

Mr. Keil	52
Mr. Dell . . .	44
Mr. Simmonds . . .	25
Mr. Mortier	10
Mr. Benchan . . .	10

Upon which the two first were declared duly elected.

The following is an extempore on the confinement of Bridewell :—

OXON, MARCH 5, 1740.

" O Bridewell ! Bridewell ! dare thy walls confine
And bar the flight of such a soul as mine?
In vain thy walls, o'er walls my soul can fly,
Scorn all thy power, and mock thy destiny.
But ah ! my body must thy force obey,
Body ! too gross to wing so light away ;
Then boast thy triumph—triumph over clay."

On Sunday evening, August 17, 1755, died, of a mortification of his foot, William Benn, Esq., Alderman of Aldersgate Ward, and President of Bridewell and Bethlem Hospitals.

Mr. Benn was chosen Alderman on December 12, 1740, on the death of Richard Levett, Esq. He was Sheriff of the City and County of Middlesex in 1742, and while he was Lord Mayor, was on January 28, 1746–47, elected President of the Bridewell and Bethlem Hospitals, in the room of Sir Robert Willimott, by a majority of 16 votes against the late Sir Daniel Lambert.

1790.—A set of villains have for some time infested the neighbourhood of Bridge Street, Ludgate Hill, who make their depredations in the most daring manner, by watching the motions of servants, of whose incautiousness and neglect they are sure to take advantage.

The house of Mr. Barnard, in Bridge Street, was robbed last week in the following manner.

Perceiving the appearance of much company in the drawing-room, one of the gang knocked at the door, and desired the maid-servant who answered it to deliver a message from a person whom he named to her master, that he would wait upon him in the morning agreeable to his desire, and requested to know the more convenient hour. Mr. B. immediately suspected the design, but before he could get downstairs the hall was stripped of a number of coats, hats, &c., with which they got clean off.

Last Saturday, May 27, 1758.—Ambrose Head, servant to Mr. W. Cox, one of the art-masters at

Bridewell, was sent with a bill to receive a large sum
of money; but not returning with it search was made
after him, and yesterday he was taken and carried
before the Right Honourable the Lord Mayor, who
committed him to the Poultry Compter, he having
embezzled £19, 11s. of the said money.

July 28, 1760.—That part of Bridewell that juts
out considerably into the street, and all the west side
of Fleet Ditch from the water-side to within two
doors of the china shop at the corner, is to be
pulled down in order to widen the passage for carriages, &c.

April 11, 1780.—This day the question to inquire
into the right of the Corporation to become Governors
of the four Royal Hospitals — St. Bartholomew's,
Christ's, Bridewell and Bethlem, and St. Thomas'—
came on at Lincoln's Inn Hall, before the Lord
Chancellor, as Visitor of all the royal foundations.

The counsel for the City of London were the
Attorney-General, the Recorder, Mr. Maddox, and
Mr. Pope.   For the petitioners (the President and
Governors), by donation, were Mr. Mansfield, Mr.
Kenyon, and Mr. Erskine.

The former, in a speech of an hour and a half,
stated the object of the petition and the prayer, and
a modern bye-law of the Corporation for sealing
Hospital leases in the Court of Common Council, that
in consequence of the new resolution, leases brought

to the Court of Aldermen agreeable to former usage, were refused the seal; after which the Lord Chancellor intimated that a matter of this importance required a deal of time, and proposed a further day convenient to the Court and Council for a complete investigation.

1800.—The following is an amusing record of eighty-eight years ago :—

"It is to be hoped that the Committee of General Purposes in the City, will order Blackfriars Bridge to be watered through the summer. As the public pay an expense towards it by a Sunday toll, they ought to be accommodated.

"At present the bridge is neither watered nor swept, and is a most intolerable nuisance."

Circa, 1792.—In one part of the building about 20 decayed artificers have houses, and about 150 boys, distinguished by white hats and blue doublets, are put apprentices to glovers, flaxdressers, weavers, &c., and when they have served their time are entitled to the freedom of the City and Ten Pounds each, towards carrying on their respective trades.[1]

"The other part of Bridewell is a prison and a house of correction for disorderly servants and vagrants, who

---

[1] Locke's, Fowke's, and Palmer's gifts were absorbed in the general fund upon the introduction of the new scheme for King Edward schools.

are made to beat hemp, and are kept at other hard
labour.

"All the affairs of the Hospital are managed by
Governors, who are above 300 in number, besides
the Lord Mayor and Court of Aldermen.

"The Governors of this Hospital are likewise Governors of Bethlem Hospital, because these two are but
one Corporation, and have also the same president,
physician, surgeon, and apothecary. Bridewell, however, has its own steward, porter, matron, and four
beadles."

There is a song by Isaac Walton, in which he im-
mortalises a certain "Old Rose." This is very likely
to have been "Rose, the old viole maker," whom
Stone mentions in his Annals as the son of John
Rose, citizen of London, living in Bridewell, and
who invented a species of lute, which he called the
"Bandon," in the fourth year of Queen Elizabeth.

In 1792 the Grand Treasury Committee investi-
gated the affairs of the Hospitals, extending their
inquiries as far back as the year 1775.

"A sum of £5957, 11s. hath been expended on
the apprentices, and £7493, 16s. 4d. in maintaining
the vagrants (the only two supposed objects of Bridewell).

"In the same period £19,254, 0s. 4d. has been
expended in salaries, &c., £6341, 6s. 1d. for taxes,
viens of estates, &c., and £3234, 9s. 1d. in feasts,

making a total of £28,829, 15s. 6½d.; and what seems still more extraordinary, the further enormous sum of £17,332, 19s. 7d. for repairs to the Hospital alone.

" Facts like these clearly demonstrate defects in the system of management, for which a radical cure should be provided."

" The average income for sixteen years past was estimated at about £4000, the average expenditure £3725, 17s. 8d.; and the Committee having hinted at the extravagance and expense and disproportions, the disbursement might be curtailed, and the saving more usefully and properly applied."

Here then we bring to a close this little volume of reminiscences of Bridewell Hospital, trusting that by its perusal, some may be moved to become Governors, and so assist and further the good work that is being done by this institution—one of the old Royal Foundations of Edward VI.

THE END.

WELLS GARDNER, DARTON, AND CO., LONDON.